Information and Instructions

This shop manual contains several sections each covering a specific group of wheel type tractors. The Tab Index on the preceding page can be used to locate the section pertaining to each group of tractors. Each section contains the necessary specifications and the brief but terse procedural data needed by a mechanic when repairing a tractor on which he has had no previous actual experience.

Within each section, the material is arranged in a systematic order beginning with an index which is followed immediately by a Table of Condensed Service Specifications. These specifications include dimensions, fits, clearances and timing instructions. Next in order of arrangement is the procedures paragraphs.

In the procedures paragraphs, the order of presentation starts with the front axle system and steering and proceeding toward the rear axle. The last paragraphs are devoted to the power take-off and power lift systems. Interspersed where needed are the additional tabular specifications pertaining to wear limits, torquing, etc.

HOW TO USE THE INDEX

Suppose you want to know the procedure for R&R (remove and reinstall) of the engine camshaft. Your first step is to look in the index under the main heading of ENGINE until you find the entry "Camshaft." Now read to the right where under the column covering the tractor you are repairing, you will find a number which indicates the beginning paragraph pertaining to the camshaft. To locate this wanted paragraph in the manual, turn the pages until the running index appearing on the top outside corner of each page contains the number you are seeking. In this paragraph you will find the information concerning the removal of the camshaft.

More information available at Clymer.com
Phone: 805-498-6703

Haynes Publishing Group
Sparkford Nr Yeovil
Somerset BA22 7JJ England

Haynes North America, Inc
859 Lawrence Drive
Newbury Park
California 91320 USA

ISBN 10: 0-87288-124-5
ISBN-13: 978-0-87288-124-2

SHOP MANUAL
MASSEY-FERGUSON

Ferguson Models TO35, F40 and TO35 Diesel
Massey-Harris Model MH50
Massey-Harris-Ferguson Model MHF202
Massey-Ferguson Models MF35, MF35D, MF50, MF202 and MF204

Tractor serial number stamped on instrument panel name plate.
Engine serial number stamped on engine name plate on side of engine.

Deluxe Model TO35

Model F40

INDEX (By Starting Paragraph)

CONDENSED SERVICE DATA

GENERAL

	Non-Diesel	Standard Diesel Standard	Perkins Diesel F. Perkins, Ltd.
Engine Make	Continental	Motor Co.	
Engine Model	Z134	23C	3A152
Number of Cylinders	4	4	3
Bore—Inches	3 5/16	3 5/16	3.6
Stroke—Inches	3 7/8	4	5
Displacement—Cu. In.	134	137.8	152.7
Compression Ratio	6.6:1-8.1:1	20:1	17,4:1
Pistons Removed From	Above	Above	Above
Main Bearings, Number of	3	3	4
Main & Rod Brgs., Adjustable?	No	No	No
Cylinder Sleeves	Wet	Dry	Dry
Forward Speeds, No. of	6	6	6
Generator & Starter Make	D-R	D-R	D-R

TUNE-UP

	Non-Diesel	Standard Diesel	Perkins Diesel
Firing Order	1-3-4-2	1-3-4-2	1-2-3
Valve Tappet Gap, Intake	0.013H	0.012C	0.010H
Valve Tappet Gap, Exhaust	0.013H	0.008C	0.010H
Inlet Valve Face Angle	30°	45°	44°
Inlet Valve Seat Angle	30°	45°	45°
Exhaust Valve Face Angle	44°	45°	44°
Exhaust Valve Seat Angle	45°	45°	45°
Ignition Distributor Make	D-R	...	...
Breaker Contact Gap	0.022	...	...
Plug Electrode Gap	0.025	...	...
Carburetor Make	‡	...	...
Float Setting, Carter	11/64	...	...
Float Setting, Marvel-Schebler	1/4	...	...
Injection Pump Make	...	C.A.V.	C.A.V.
Injection Pump Model	...	D.P.A.	D.P.A.
Injection Pump Timing	...	**	18°BTC
Injector Nozzle Make	...	C.A.V.	C.A.V.
Injector Nozzle Setting	...	1900 Psi	1760 Psi
Engine Low Idle Rpm	475	500	500
Engine High Idle Rpm	2200	2200	2200
Engine Loaded Rpm	2000	2000	2000
Pto High Idle Rpm	——— 792 @ 2200 ———		
Pto Loaded Rpm	——— 720 @ 2000 ———		
Pto Rated Rpm	——— 540 @ 1500 ———		

SIZES—CAPACITIES—CLEARANCES
(Clearances in thousandths)

Crankshaft Journal Diameter	2.2495	2.7515	2.7485
Crankpin Diameter	1.937	2.311	2.2485
Camshaft Journal Diameter:			
Front	1.8092	1.559	1.869
Center	1.7462	1.684	1.859
Rear	1.6837	1.684	1.839
Piston Pin Diameter	0.8592	1.1249	1.250
Valve Stem Diameter:			
Inlet	0.3145	0.3109	0.311
Exhaust	0.3128	0.3729	0.311
Main Bearing, Diameter Clearance	1.5-2.5	2-3.5	2.5-4.5
Rod Bearing, Diameter Clearance	1.5-2.5	2-3.5	2-3.5
Piston Skirt Clearnce	1.2	3.5-4.4	...
Crankshaft End Play	2-6	5-12	2-11
Camshaft Brg., Dia. Clearance:			
Front	2.5-4.5	2.5-4.5	4-8
Center	2.5-4.5	1-3.3	4-8
Rear	2.5-4.5	1-3.3	4-8
Camshaft End Play	3-7	2-7.5	...
Cooling System—Gallons	2.5	2.25	2.25
Crankcase Oil—Quarts	5 *	7 *	7 *
Transmission and Differential—Qts.	32†	32†	32†
Belt Pulley—Quarts	1	1	1

*One additional quart when oil filter is renewed.

**17°BTDC on engine equipped with DPA3240011 Pump (without automatic advance); 13°BTDC on engine equipped with DPA-3242645 pump (with automatic advance).

†Includes hydraulic system. If auxiliary hydraulic equipment is used where the auxiliary equipment oil capacity exceeds 1½ gallons, additional oil must be added; then, the additional oil should be drained when the auxiliary equipment is removed. On Model MF204, "Revers-O-Matic" unit contains 10 qts. of automatic transmission fluid, Type "A", in addition.

‡Carter or Marvel-Schebler.

Model MH50

Model MHF202

Model MF50 LP Gas

Model TO35 Diesel (Standard Engine)

FRONT SYSTEM (AXLE TYPE)

AXLE ASSEMBLY

Models TO35-TO35D-MF35-MF35D

1. The front axle main (center) member, complete with extensions and wheels, can be removed as follows: Drain cooling system and remove hood assembly and radiator. Disconnect both radius rods from side of transmission case and both drag links from pitman arms. Support tractor, remove the axle pivot pin retaining screw and withdraw pivot pin from axle support.

Reject diameter for the pivot pin bushing (13—Fig. MF225) is 1.790. When installing new bushing, make certain that split side is up and ream the bushing, after installation, to an inside diameter of 1.7615-1.7675. Diameter of new axle pivot pin (10) is 1.747-1.748. Reject diameter for the pivot pin is 1.700.

Note: Axle pivot pin and/or bushing should not be lubricated. Squeaking at this point can be eliminated by using dry graphite.

When installing axle assembly, reverse the removal procedure. Shims, located under radius rod ball caps, can be removed to compensate for excessive wear at that point; but, sufficient shims should remain to eliminate any possibility of binding.

Models F40-MH50-MF50

2. To remove the axle main (center) member (11—Fig. MF226), support tractor under engine, remove the grille lower panel, disconnect tie rods from the spindle steering arms and remove both axle extensions and wheel assemblies from the center member. Unbolt and remove the axle pivot thrust plate (16) and save shims (15) for reinstallation. Remove the cap screws retaining pivot bracket (13) to the front support casting and remove the pivot bracket. Pull center member (11) forward and out of rear pivot bushing (12) which is located in the front support casting.

Inside diameter of new pivot bushings is 1.877-1.879 for rear bushing (12) and 2.002-2.004 for front bushing (14). Pivot pin diameter for a new center member is 1.874-1.876 for the rear pin and 1.999-2.001 for the front pin. Renew center member and/or bushings if running clearance is excessive. Replacement bushings are **pre-sized and will not require reaming**

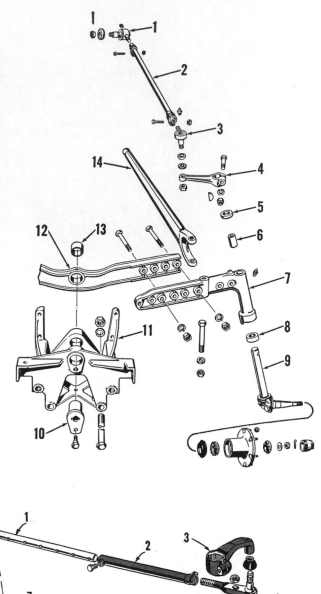

Fig. MF225 — Exploded view of models TO35 & MF35 front axle and associated parts. Recommended front wheel toe-in of 0-¼ inch is obtained by varying the length of each drag link on equal amount.

1. Drag link end, rear
2. Drag link
3. Drag link end, front
4. Spindle steering arm
5. Dust seal
6. Bushing
7. Axle extension
8. Bearing
9. Spindle
10. Axle pivot pin
11. Front axle support
12. Axle main member
13. Pivot bushing
14. Radius rod

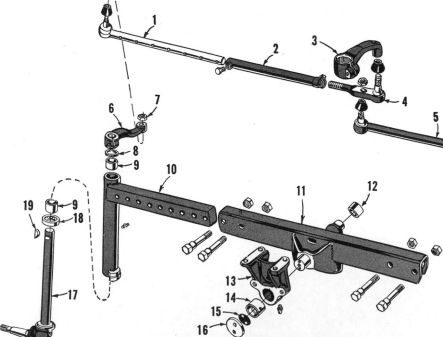

Fig. MF226—Utility and Hi-arch front axle used on models F40, MF50 and MF50. Recommended axle end play of 0.002-0.008 is adjusted with shims (15). Toe-in adjustment is accomplished by turning tie rod tube (2).

1. Right tie rod
2. Tie rod tube
3. Steering crank arm
4. Tie rod end
5. Left tie rod tube
6. Spindle steering arm
7. Nut
8. Dust seal
9. Spindle bushings
10. Axle extension
11. Axle center member
12. Rear pivot bushing
13. Pivot bracket
14. Front pivot bushing
15. Shims
16. Thrust plate
17. Spindle and knuckle
18. Thrust bearing
19. Woodruff key

if carefully installed with a suitable arbor. Make certain, however, that lubrication holes in bushings are in register with similar holes in pivot bracket and front support.

When installing the axle center member, reverse the removal procedure and vary the number of shims (15) to obtain a center member end play of 0.002-0.008 when checked between rear face of pivot bracket (13) and front face of center member (11). Shims are available in thicknesses of 0.002, 0.005 and 0.010.

Models MHF202-MF202-MF204

3. The front axle main member complete with spindles and wheels can be removed as follows: Drain cooling system and remove hood assembly and radiator. On Model MF204 it will first be necessary to remove oil cooling radiator and oil lines. Disconnect both drag links and both power steering cylinders from the spindle steering arms. Disconnect both power steering cylinders from the axle main member and, without disconnecting the oil lines, lay the cylinders rearward and out of way. Unbolt and remove thrust plate from axle support and withdraw spacer (S—Fig. MF228) and shims. Remove bolt (B) retaining axle to pivot pin.

Support tractor under oil pan, unbolt front support casting from engine and roll axle and support assembly away from tractor. Bump the axle pivot pin forward and out of axle and support casting.

Inside diameter of new pivot bushings is 1.8765-1.8795. Diameter of new pivot pin is 1.8745-1.8750. Renew pivot pin and/or bushings if clearance is excessive. Replacement bushings are pre-

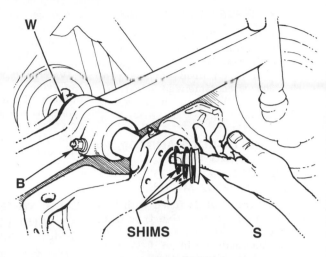

Fig. MF228 — Installing shims which control end play of models MHF202, MF202 and MF204 axle pivot pin.

sized and will not require reaming if carefully installed with a suitable arbor. Make certain, however, that lubrication holes in bushings are in register with similar holes in the support casting.

When installing the axle main member, reverse the removal procedure and be sure to install thrust washer (W) between rear face of axle and the support casting. Tighten bolt (B) securely; then vary the number of shims at front of pivot pin to obtain an axle and pivot shaft end play of 0.002-0.008 when checked between rear face of axle and the front face of the thrust washer.

SPINDLE BUSHINGS
Models TO35-MF35

4. Each axle extension (support) contains two renewable, split type bushings which require final sizing after installation to an inside diameter of 1.249-1.250. Recommended clearance between spindles and bushings is 0.0035-0.005.

Models F40-MH50-MF50

5. Each axle extension contains two renewable bushings which require final sizing after installation to an inside diameter of 1.249-1.250. Recommended clearance between spindles and bushings is 0.0035-0.005.

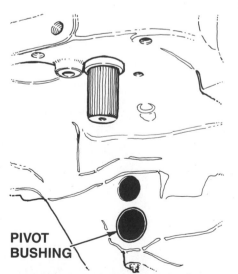

PIVOT BUSHING

Fig. MF227—Axle pivot pin rear bushing installation on F40, MH50 and MF50.

Fig. MF229 — Exploded view of the single wheel fork, support casting and associated parts used on models F40, MH50 and MF50. End play of wheel fork is controled by shim washers (13).

1. Link
2. Steering arm
3. Steering arm
4. Wheel fork
5. Oil seal
6. Bearing race
7. Needle thrust bearing
8. Bearing race
9. Lower needle bearing
10. Oil seal
11. Oil seal
12. Upper needle bearing
13. Shim washers
14. Thrust plate
15. Lock washers
16. Cap screws
17. Dust cap

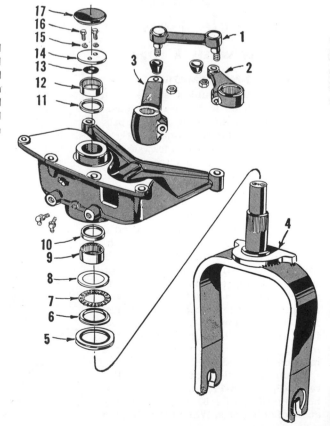

Models MHF202-MF202-MF204

6. Each end of the front axle main member contains two renewable bushings which require final sizing after installation to an inside diameter of 1.499-1.500. Recommended clearance between spindles and bushings is 0.0035-0.005.

TOE-IN, TIE-RODS AND/OR DRAG LINKS
Models TO35-MF35 MHF202-MF202-MF204

7. Each drag link is fitted with two non-adjustable, automotive type drag link ends.

Recommended toe-in of 0-¼ inch is adjusted by varying the length of each drag link an equal amount.

Models F40-MH50-MF50

8. Tie-rod ends are of the non-adjustable, automotive type.

Recommended toe-in of 0-¼ inch is adjusted by varying the length of the right hand tie rod. This is accomplished by loosening the tube set screw and clamp bolt and turning the tube (2—Fig. MF226) either way as required. Be sure the tube set screw is in the forward position before tightening. Center to center distance between drag link ends is 49⅝ inches.

FRONT SYSTEM (Tricycle Type)

SINGLE WHEEL & FORK

Models F40-MH50-MF50

9. The fork mounted single front wheel is carried in taper roller bearings which should be adjusted to provide a very slight rotational drag.

To remove the wheel fork, support tractor under engine and remove the grille lower panel. Open the grille door and remove the sheet metal dust cap (17—Fig. MF229). Remove the cap screws (16) retaining thrust plate (14) to upper end of wheel fork and withdraw thrust plate (14) and shim washers (13). Working through opening in the support casting, loosen the clamp bolt retaining steering arm (3) to wheel fork. Raise front of tractor and at the same time, withdraw the wheel fork from below. CAUTION: Wheel fork must not be cocked during removal or seals and needle bearings may be damaged.

Examine needle bearings (9 and 12), seals (5, 10 and 11) and needle thrust bearing (6, 7 and 8) and renew any questionable parts. When installing the caged needle bearings (9 and 12), be sure to align the oil hole in the bearings with the oil feed holes in the support casting.

Install the wheel fork by reversing the removal procedure and vary the number of shim washers (13) to provide the wheel fork with an up and down end play of 0.002-0.008.

Note: The wheel fork and steering arm (3) have a blind spline to facilitate correct installation.

DUAL WHEELS, LOWER SPINDLE AND LOWER PEDESTAL
Models F40-MH50-MF50

10. Each of the dual wheels is mounted on taper roller bearings which should be adjusted to provide a very slight rotational drag.

To remove the lower spindle and wheels assembly, support tractor under engine and remove the grille lower panel. Open the grille door, unlock and remove cap screw (16—Fig. MF230), locking washer (15) and flat washer (14). Working through opening in the support casting, loosen the clamp bolt retaining steering arm

(13) to the spindle. Raise front of tractor and at the same time, withdraw the spindle, axle and wheels assembly from below.

Lower pedestal (6) can be removed from the support casting at this time.

Examine bearing cones (4 and 8), cups (5 and 7) and seals (3 and 9) and renew any questionable parts.

Install the spindle and lower pedestal by reversing the removal procedure and tighten cap screw (16) to remove all spindle end play without causing any bearing drag.

Note: The lower spindle (12) and steering arm (13) have a blind spline to facilitate correct installation.

MANUAL STEERING SYSTEM

Models TO35-MF35

11. **ADJUSTMENT.** Before attempting to adjust the steering gear, first make certain that the gear housing is properly filled with lubricant, then disconnect drag links from pitman arms to remove load from gear unit.

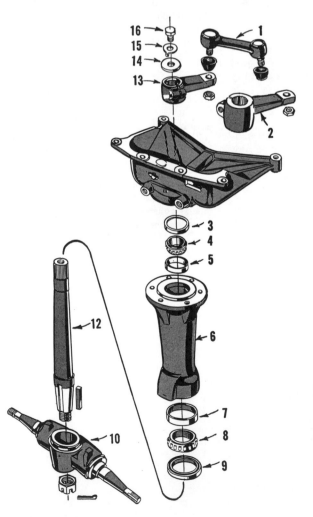

Fig. MF230 — Exploded view of the dual wheel tricycle lower spindle, pedestal and related parts. Spindle bearings should be adjusted to remove all end play without binding.

1. Link
2. Steering arm
3. Oil seal
4. Bearing cone
5. Bearing cup
6. Lower pedestal
7. Bearing cup
8. Bearing cone
9. Oil seal
10. Axle
12. Lower spindle
13. Steering arm
14. Washer
15. Locking washer
16. Cap screw

12. WORM SHAFT BEARINGS. To check and/or adjust the steering worm shaft bearings, first loosen the lock nuts and back off the right and left hand sector adjusting screws (31 Fig. MF232) approximately four turns. Attach a spring scale to outer edge of steering wheel and check the amount of pull required to keep the steering wheel in motion after it has crossed the mid or center point of its rotation. The wormshaft bearings are properly pre-loaded when the spring scale reading is ½-1½ pounds. If bearing adjustment is not as specified, unbolt steering housing cover from housing and vary the number of shims (26) until the desired bearing pre-load is obtained. Shims are available in thicknesses of 0.002, 0.005, 0.010 and 0.030.

13. SECTOR BACKLASH. With the wormshaft bearings properly adjusted as outlined in paragraph 12, proceed as follows: Turn the steering wheel to the mid (or straight ahead) position (lower ends of Pitman arms over radius rod ball joints) and using a screw driver, turn the adjusting screw for the left hand sector shaft until a spring scale pull of 1¾-3 pounds (measured at rim of steering wheel) will keep the wheel in motion after it has crossed the mid or center point of its rotation.

With the left sector shaft adjusted, turn the adjusting screw for the right sector shaft until a spring scale pull of 3¼-5 pounds (measured at rim of steering wheel) will keep the steering wheel in motion after it has crossed the mid or center point of its rotation.

14. REMOVE AND REINSTALL. To remove the steering gear housing, instrument panel and transmission cover assembly, proceed as follows: Tip hood assembly forward, drain cooling system and disconnect the heat indicator sending unit from the water outlet elbow. Disconnect cable from starting motor and wires from coil, headlights and generator. Disconnect tractormeter cable from generator, choke rod from carburetor and oil gage line from right side of cylinder block. On Diesel models, disconnect tractormeter cable, fuel shut-off control rod and throttle control rod from injection pump; then disconnect oil pressure banjo fitting at left side of block. Remove air cleaner pipe, shut off fuel and remove fuel line. Unbolt fuel tank from its rear support, loosen the fuel tank front support bolts and block up between the fuel tank and rocker arm cover. Loosen the U-bolt assembly from front end of throttle rod. Disconnect battery cables and remove battery. Disconnect tail light wires and wires from starter safety switch. Disconnect drag links from pitman arms, unbolt battery platform from engine, unbolt transmission cover from transmission

and lift the steering gear and housing assembly from tractor.

15. OVERHAUL. The steering gear unit can be overhauled without removing gear housing assembly from tractor by removing the battery and the instrument panel and battery platform assembly as outlined in paragraph 15A. Some mechanics prefer to remove the complete gear housing assembly as outlined in paragraph 14 before attempting to disassemble the steering gear. In either case, however, the time involved is about the same.

15A. To remove the instrument panel, tip hood assembly forward, drain cooling system and disconnect the heat indicator sending unit from the water outlet elbow. Disconnect cable from starting motor and wires from coil, headlights and generator. Disconnect tractormeter cable from generator and choke rod from carburetor. On Diesel models, disconnect tractormeter cable, fuel shut-off control rod and throttle control rod from injection pump. Remove air cleaner and rubber intake tube and disconnect oil line from gage. Shut off fuel and remove fuel line. Unbolt fuel tank from its rear support, loosen the fuel tank front support bolts and block-up between the fuel tank and the rocker arm cover. Loosen the U-bolt assembly from front end of throttle rod. Disconnect battery cables and remove battery. Disconnect tail light wires and starter safety switch wires. Remove the steering wheel nut and using a suitable puller, remove steering wheel from shaft. Remove the steering column felt seal, cap and spring. Unbolt and remove battery platform and instrument panel assembly from tractor.

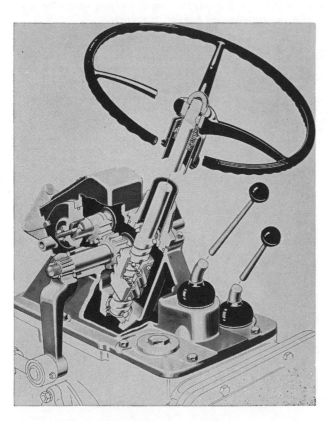

Fig. MF231 — Cutaway view of models TO35 and MF35 manual steering gear. The steering gear housing and transmission top cover are an integral unit. The steering gear is provided with a shim adjustment for the wormshaft bearings and screw adjustments for backlash of both sectors.

Fig. MF232—Models TO35 & MF35 steering gear adjustments. Shims (26) control wormshaft bearing adjustment. Screws (31) control the sector backlash.

19. Steering column & housing top cover
28. Steering housing side cover

16. Disconnect drag links from pitman arms and unbolt and pull pitman arms from sector shafts. Unbolt the side cover for the upper sector shaft (36—Fig. MF233) and remove lock nut from screw (31). Using a screw driver, turn the adjusting screw in and remove the side cover and sector shaft. Remove the other side cover and sector shaft in the same manner. Unbolt steering housing upper cover from housing and remove cover, shaft and ball nut assembly. Do not disassemble the ball nut assembly (41) as component replacement parts are not available. If the steering shaft and/or ball nut are damaged, renew the complete assembly. The need and procedure for further disassembly and/or overhaul is evident.

Shims (30) on the adjusting screws (31) are available in thicknesses of 0.063, 0.065, 0.067 and 0.069. When reassembling, use the proper combination of shims to provide a very slight amount of clearance between adjusting screw head and slot in sector shafts.

When reassembling, install the steering shaft and ball nut assembly and bolt the housing upper cover assembly in position, using the necessary number of shims (26) to very slightly pre-load the shaft bearings. Turn the steering shaft until the ball nut raises to a position where the lower gear tooth on the ball nut is clearly visible through the right side opening in housing. Hold the lower sector shaft (the one with the greater number of teeth) and side cover assembly with blank portion of gear down and install shaft through right side opening in housing. Mesh sector gear with ball nut so that lower tooth on ball nut meshes with the groove next to the blank portion of sector shaft gear. Install the upper sector shaft, meshing the double punch marked tooth space on upper sector gear with the single punch marked tooth on the lower sector gear. Bolt side covers to steering gear housing and install the adjusting screw locknuts.

Turn steering gear to its mid or straight ahead position and install pitman arms (longer one on right side) so that drag link attaching holes are directly above the radius rod ball joints.

When installation is complete, fill gear housing with lubricant and adjust the unit as outlined in paragraphs 11, 12 and 13.

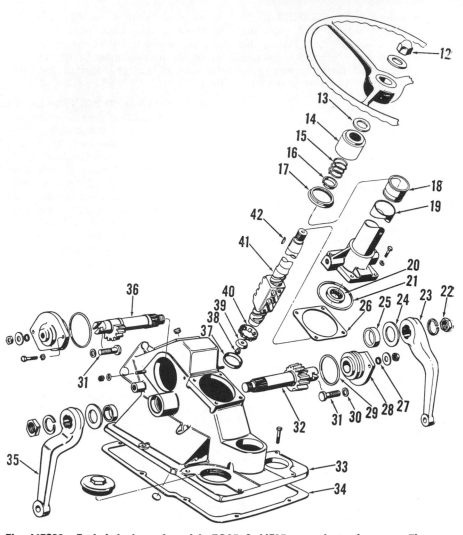

Fig. MF233—Exploded view of models TO35 & MF35 manual steering gear. The gear housing is also the transmission housing top cover.

12. Nut	20. Roller bearing & cage	27. Packing	34. Gasket
13. Felt seal	21. "O" ring packing	28. Housing side cover	35. Pitman arm, left
14. Steering column cap	22. Pitman arm nut	29. "O" ring packing	36. Sector shaft, single
15. Spring	23. Pitman arm, right	30. Shim	37. Bearing cup
16. Spring seat	24. Dust seal	31. Lash adjusting screw	38. Eyelet
17. Cap seal	25. Oil seal	32. Sector shaft, double	39. Bearing retainer
18. Bearing	26. Shims	33. Housing	40. Roller bearing & cage
19. Steering column and housing top cover			41. Steering shaft & ball nut assembly
			42. Key

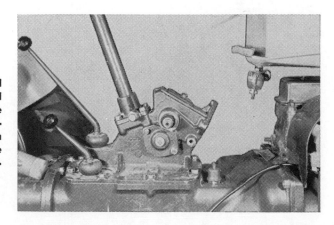

Fig. MF234 — The model TO35 & MF35 manual steering gear unit can be overhauled without removing the assembly from tractor by removing the instrument panel and battery platform assembly.

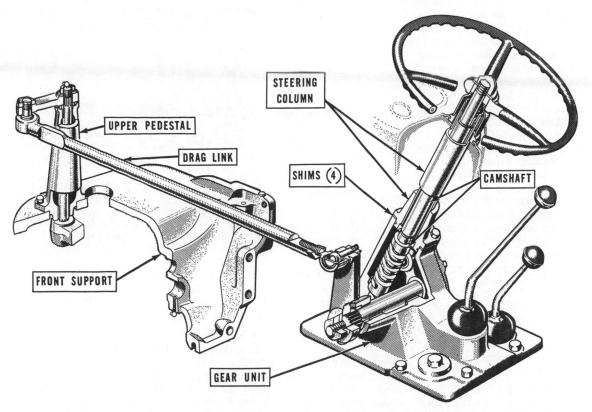

Fig. MF235—Cut-away view of the cam and lever type steering gear unit used on F40, MH50 and MF50. Pedestal unit shown in this illustration is used on models with manual steering only. The gear unit, however, is the same on models with or without power steering.

Models F40-MH50-MF50

For the purposes of this section, the F40, MH50 and MF50 manual steering system will include the cam and lever type gear unit, the housing of which is integral with the transmission top cover; and the upper pedestal assembly which is mounted on the tractor front support. Refer to Fig. MF235.

17. ADJUST GEAR UNIT. Before attempting to adjust the steering gear unit, first make certain that the gear housing is properly filled with lubricant, then disconnect the drag link from the pitman arm to remove load from the gear unit.

18. CAMSHAFT BEARINGS. To check and/or adjust the camshaft bearings, first loosen the adjusting screw locknut (17—Fig. MF236) and back off the lever shaft adjusting screw (16) at least two full turns. Pull up and push down on the steering wheel to check for camshaft end play. If end play exists, unbolt the steering column from gear housing and raise the column; then, split and remove sufficient quantity of shims (4—Fig. MF235) to allow camshaft to turn with a barely perceptible drag. Shims are available in thicknesses of 0.002, 0.003 and 0.010.

19. LEVER BACKLASH. With the camshaft bearings adjusted as outlined in the preceding paragraph, turn the steering gear to the mid or straight ahead position (pitman arm in vertical postion) and turn adjusting screw (16—Fig. MF236) in to obtain a very slight drag when steering gear is rotated through this mid-position. The gear unit should turn freely in all other positions. Tighten lock nut (17) when adjustment is complete.

20. **REMOVE AND REINSTALL GEAR UNIT.** To remove the steering gear housing assembly, drain cooling system, remove the hood, side panels and grille assembly and disconnect the heat indicator sending unit from the water outlet elbow. Disconnect cable from starting motor and wires from coil, headlights and generator. Disconnect tractormeter cable from generator, choke rod from carburetor and oil gage line from right side of cylinder

Fig. MF236—On models F40, MF50 & MF50, steering gear lever shaft should be adjusted to provide a slight drag when gear unit is rotated through the mid position.

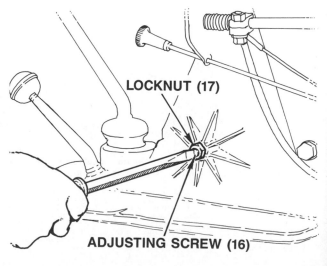

LOCKNUT (17)

ADJUSTING SCREW (16)

block. Shut off the fuel and remove the fuel line. Unbolt the fuel tank from its rear support, loosen the fuel tank front support bolts and block up between fuel tank and rocker cover.

Note: Some mechanics prefer to completely remove the fuel tank. Loosen the U-bolt assembly from front end of throttle rod. Disconnect battery cables and remove battery. Disconnect tail light wires and wires from starter safety switch. Unbolt battery platform from engine and remove steering wheel, steering wheel Woodruff key, felt washer (25—Fig. MF237), spring (22), chrome cap (24) and rubber seal (23). Remove the cap screws retaining the instrument panel and battery platform to steering gear housing and lift the instrument panel and battery platform assembly from tractor.

21. Disconnect drag link from pitman arm, remove the cap screws retaining the transmission cover to transmission housing and lift the steering gear and transmission cover unit from tractor.

22. OVERHAUL GEAR UNIT. The steering gear unit can be overhauled

Fig. MF238—Upper pedestal and drag link installation on models F40, MH50 & MF50 equipped with manual steering and adjustable type front axle. The installation on tricycle models is similar except for details of the lower steering arm.

without removing the gear housing assembly from tractor by removing the instrument panel and battery platform assembly as outlined in paragraph 20.

Remove pitman arm from lever shaft and remove the gear housing side cover (18—Fig. MF237). Withdraw the lever shaft and stud assembly. Examine the stud and renew same if damaged or excessively worn. Note:

On axle type tractors, the stud (14A) is mounted in roller bearings, but component parts of the stud assembly are not sold separately.

Remove the steering column and gear housing cover (2) and save shims (4) for reinstallation. Withdraw the camshaft, remove snap rings (5), cups (6) and bearing balls (7). Examine all parts and renew any which are questionable.

New lever shaft bushings (11) should be reamed after installation, if necessary, to provide an inside diameter of 1.6235-1.625. The lever shaft (15) should have a clearance of 0.0005-0.003 in the bushings.

When reassembling the unit, adjust the camshaft bearings as in paragraph 18 and the lever backlash as in paragraph 19.

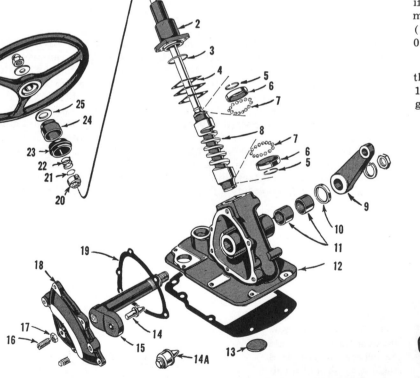

Fig. MF237—Exploded view of the cam and lever type steering gear unit used on models F40, MH50 and MF50. Camshaft bearings and lever backlash are adjustable.

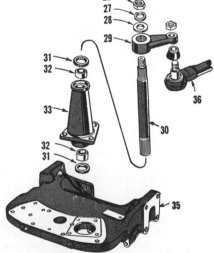

Fig. MF239—Exploded view of the upper pedestal and components used on models F40, MH50 and MF50 with manual steering.

1. Woodruff key	8. Cam and shaft assembly	14. Lever stud (tricycle models)
2. Steering column and cover	9. Pitman arm	14A. Lever stud (axle models)
3. "O" ring	10. Oil seal	15. Lever shaft
4. Shims	11. Bushings	16. Adjusting screw
5. Snap ring	12. Housing	17. Lock nut
6. Ball cup	13. Housing plug	
7. Bearing balls		

18. Side cover	
19. Gasket	
20. Bearing assembly	
21. Bearing spring seat	
22. Bearing spring	
23. Seal	
24. Steering column cap	
25. Felt seal	

26. Nut	30. Pedestal shaft
27. Lock washer	31. Dust seals
28. Flat washer	32. Bushings
29. Upper steering arm	33. Pedestal
	35. Front support

23. R&R AND OVERHAUL UPPER PEDESTAL. To remove the upper pedestal, first remove the grille screens and the left side panel as shown in Fig. MF238. Loosen clamp bolt (B) retaining steering arm to lower end of pedestal shaft and disconnect the drag link at forward end. Remove the cap screws retaining upper pedestal to front support casting and remove pedestal assembly from tractor.

Remove the nut (26 - Fig. MF239) retaining steering arm (29) to pedestal shaft and remove the steering arm. Remove pedestal shaft (30), examine all parts and renew any which are damaged or show wear. Ream new bushings (32), after installation, to an inside diameter of 1.5005-1.5015. Pedestal shaft (30) should have a clearance of 0.0005-0.002 in the bushings.

POWER STEERING SYSTEM

Note: The maintenance of absolute cleanliness of all parts is of utmost importance in the operation and servicing of the hydraulic power steering system. Of equal importance is the avoidance of nicks or burrs on any of the working parts.

FILLING AND BLEEDING
All Models So Equipped

24. Fluid capacity for the complete power steering system is ⅔ qt. for TO35, MF35 and MF50 and 1 qt. for other models.

Only automatic transmission fluid, Type A is recommended for use in the power steering system. Reservoir fluid level and the paper filter element in reservoir should be checked every 750 hours of operation or more often in severe dust conditions. Reservoir fluid level should be maintained ¼-½-inch above the filter element.

To bleed the system, fill reservoir, start engine and turn the steering wheel full right and full left several times to bleed air from the system; then, refill reservoir to the proper level.

TROUBLE SHOOTING
All Models So Equipped

25. The accompanying table lists troubles which may be encountered in the operation of the power steering system. The procedure for correcting most of the troubles is evident; for those not readily remedied, refer to the appropriate subsequent paragraphs.

NOTE: Control valve on F40, MH50 and MF50 is not equipped with check valve. Turning steering wheel fully with engine not running will exhaust fluid in cylinder causing pressure build-up in reservoir which may rupture cover seal causing fluid loss. If this condition is encountered, caution operator against operating steering with engine not running.

POWER STEERING SYSTEM TROUBLE-SHOOTING CHART

	Loss of Power Assistance	Power Assistance in One Direction Only	Unequal Turning Radius	Erratic Steering Control	Fluid Foaming Out of Reservoir
Binding, worn or bent mechanical linkage	★		★	★	
Insufficient fluid in reservoir	★				
Low pump pressure	★				
Faulty or improperly installed control valve thrust bearing (TO35, MF35, MHF202, MF202 & MF204	★	★		★	
Valve thrust bearing nut improperly tightened (TO35, MF35, MHF202, MF202 & MF204	★			★	
Sticking or binding valve spool (TO35, MF35, MHF202, MF202 & MF204	★	★		★	
Faulty valve plungers and/or springs (TO35, MF35, MHF202, MF202 & MF204	★			★	
Check valve ball not seating (TO35, MF35, MHF202, MF202 & MF204	★				★
Damaged or restricted hose or tubing	★	★			★
Pump to valve hose lines reversed	★				
Wrong fluid in system	★			★	★
Improperly adjusted tie rods or drag links			★	★	
Steering arms not positioned properly			★		
Air in system				★	★
Plugged filter element					★
Internal leak in valve					★
Faulty cylinder	★			★	
Faulty linkage adjustment (F40, MH50, MF50)				★	

SYSTEM OPERATING PRESSURE AND RELIEF VALVE

All Models So Equipped

26. A pressure test of the hydraulic circuit will disclose whether the pump, relief valve or some other unit in the system is malfunctioning. To make such a test, proceed as follows:

Connect a pressure test gage and shut-off valve in series with the pump pressure line as shown in Fig. MF240. Note that the pressure gage is connected in the circuit between the shut-off valve and the pump. Open the shut-off valve and run engine at low idle speed until oil is warmed. Advance the engine speed to 2,000 rpm, close the shut-off valve and retain in the closed position only long enough to observe the gage reading. Pump may be seriously damaged if valve is left in the closed position for more than 10 seconds. If gage reading is 1,100 psi with the shut-off valve closed, the pump and relief valve are O. K. and any trouble is located in the control valve, power cylinder and/or connections.

If the pump output pressure is more than 1,100 psi, the relief valve is either improperly adjusted or stuck in the closed position. If the output pressure is less than 1,100 psi, either the relief valve is improperly adjusted or the pump requires overhauling. In any event, the first step in eliminating trouble is to adjust the relief valve. This may be accomplished by removing the reservoir cover and filter element and turning the relief valve adjusting screw (20—Fig. MF241) either way as required. One turn of the adjusting screw will increase the output pressure approximately 300 psi. Be sure to stake the screw after adjustment is complete. If relief valve

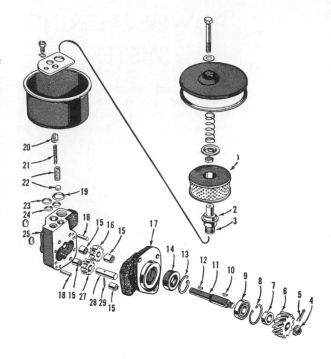

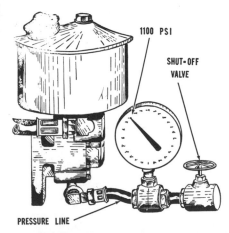

Fig. MF240 — Pressure gage and shut-off valves installation for checking the power steering system operating pressure. At 2000 engine rpm, the pump output pressure should be 1100 psi.

Fig. MF241 — Exploded view of the power steering pump used on all models. The pump is driven from the engine timing gear train and delivers approximately 4 gallons per minute at an engine speed of 2000 rpm.

1. Filter element
2. Element seat cup
3. Stud
4. Nut
5. Cotter pin
6. Drive gear
7. Bearing spacer
8. Snap ring
9. Ball bearing
10. Woodruff key
11. Drive shaft
12. Woodruff key
13. Snap ring
14. Oil seal
15. Needle bearings
16. Driven gear
17. Pump body
18. Dowel pin
19. "O" ring
20. Adjusting screw
21. Spring
22. Relief valve spring and ball
23. "O" ring
24. "O" ring
25. Gear housing
27. Follower gear
28. Follower gear shaft
29. Shear pin

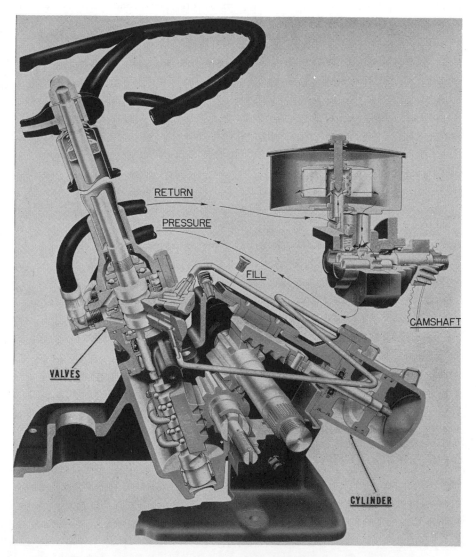

Fig. MF242—Cut-away view of TO35 & MF35 power steering system.

adjustment will not restore the pressure with the shut-off valve closed, overhaul the pump as in paragraph 27.

PUMP
All Models So Equipped

27. **OVERHAUL.** With the pump removed from engine, remove the reservoir, drive gear (6—Fig. MF241) and Woodruff key (10). Separate the gear housing (25) from pump body (17) and remove the pumping gears (16 and 27) and Woodruff key (12). Remove snap ring (8) and extract drive shaft (11). The need and procedure for further disassembly is evident.

Examine all parts and renew any which are questionable. When reassembling, be sure to renew all "O" rings and seals.

After pump is installed on engine, check the system operating pressure as in paragraph 26.

STEERING VALVES
Models TO35-MF35-MHF202-MF202-MF204

28. **R&R AND OVERHAUL.** To remove the power steering valves, first remove the instrument panel assembly as outlined in paragraph 15A.

Disconnect lines and tubing from the valve body, remove the three cap screws retaining the steering column to the gear unit and lift off the steering column. Refer to Fig. MF244. Unstake and remove lock nut (5), but do not allow the steering shaft to turn when removing the nut. Remove the Belleville washer (6) and upper thrust bearing (7); then lift off the valve body and spool assembly and the lower thrust bearing (9). Unbolt and remove the adapter (16) from steering gear housing, but do not allow the steering shaft to move upward.

Remove the valve centering plungers (20) and centering springs

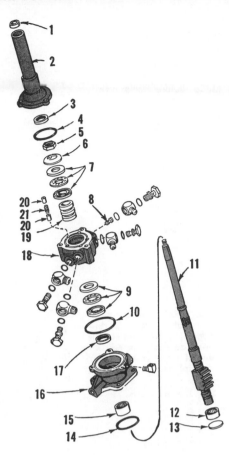

(21) from valve body and push spool (19) from the body bore. Thoroughly clean and examine all parts for damage or wear. The valve spool and body are mated parts and must be renewed as an assembly. The plungers and springs are available separately. The check valve in the return port of valve body can be removed with a screw driver and must be renewed as an assembly. Be sure to check the condition of needle bearing (15) and seal (17) in the adapter casting.

When reassembling, be sure to lubricate all parts in Automatic Transmission Fluid, Type A. Install the adapter casting and tighten the retaining screws securely. Place the smaller washer of lower thrust bearing (9)

Fig. MF244—Exploded view of TO35 & MF-35 power steering valve and associated parts. Models MHF202, MF202 and MF204 are similar.

1. Upper bearing	12. Needle bearing
2. Steering column	13. Expansion plug
3. Seal	14. "O" ring
4. "O" ring	15. Needle bearing
5. Lock nut	16. Adapter
6. Belleville washer	17. Oil seal
7. Upper thrust bearing	18. Valve body
8. Check valve	19. Valve spool
9. Lower thrust bearing	20. Valve centering plunger
10. "O" ring	21. Valve centering spring
11. Steering shaft assembly	

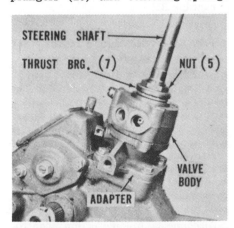

Fig. MF243—View showing TO35 & MF35 power steering valve installation after steering column is removed. Models MHF202, MF202 and MF204 are similar.

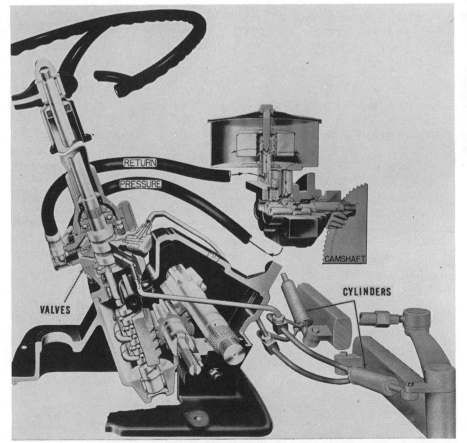

Fig. MF245—Cut-away view of models MHF202, MF202 and MF204 power steering system. The steering cylinders are connected directly to the spindle steering arms.

in the adapter casting with ball groove up; then install the ball bearing and the larger washer, ball groove down. Assemble the centering plungers (20) and spring (21) in the valve body and install spool (19) so that the identifying groove in I.D. of spool is toward the same side of body as the port identification symbols "PR" and "RT". Then install the assembled valve body with the symbols "PR" and "RT" up. Install the upper thrust bearing (7) with larger diameter washer toward the valve body. Install Belleville washer (6) with convex side up. Install lock nut (5) and while preventing the steering shaft from turning, tighten the lock nut to a torque of 20-30 Ft.-Lbs.; then back off the nut ¼ turn and stake in place. Assemble the remaining parts and bleed the system as in paragraph 24.

Models F40-MH50-MF50
Before Ser. No. 516788

29. R&R AND OVERHAUL. To remove the power steering valve, open the grille door and disconnect the hose lines from the valve unit. Remove pin (P—Fig. MF246), unbolt and remove the valve unit from cylinder.

To disassemble the unit, disconnect the adjusting linkage from link (10—Fig. MF247) and pry out the dust seal retainer (6). Extract the dust seal (7), remove snap ring (8) and withdraw

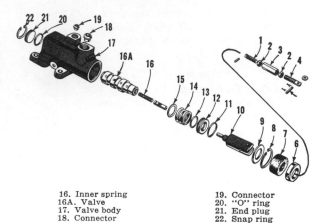

Fig. MF247 — Exploded view of the power steering valve unit used on models F40, MH50 and MF50 before serial No. 516788.

1. Valve link end
2. Jam nut
3. Adjusting sleeve
4. Link end
6. Dust seal retainer
7. Dust seal
8. Snap ring
9. Washer
10. Valve link
11. "O" ring
12. Valve guide
13. "O" ring
14. Outer spring
15. Washer

16. Inner spring
16A. Valve
17. Valve body
18. Connector

19. Connector
20. "O" ring
21. End plug
22. Snap ring

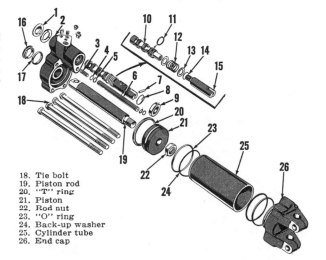

Fig. MF248 — Exploded view of power steering control valve and cylinder assembly used on model MF50 tractor after serial number 516787.

1. Valve cap
2. Valve housing
3. Spring
4. "O" ring
5. Back-up washer
6. Oil tube
7. Roll pin
8. Snap ring
9. Oil seal
10. Control valve
11. "O" ring
12. Centering spring
13. Centering washer
14. "O" ring
15. Valve rod
16. Dust seal
17. "T" ring

18. Tie bolt
19. Piston rod
20. "T" ring
21. Piston
22. Rod nut
23. "O" ring
24. Back-up washer
25. Cylinder tube
26. End cap

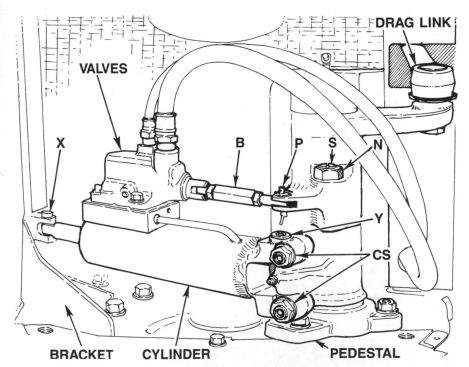

Fig. MF246—Power steering valves, cylinder and pedestal installation on models F40, MH50 and MF50. The power steering system incorporates the same cam and lever steering gear as models with manual steering.

the valve components from the body. Thoroughly clean and renew any questionable parts. When reassembling, be sure to renew all "O" rings and seals and install a new seal retainer (6). After control valve is installed, bleed system as outlined in paragraph 24 and adjust linkage as outlined in paragraph 31.

MF50 After Ser. No. 516787

30. R&R AND OVERHAUL. The power steering control valve body is cast into the power cylinder rod end housing. To remove the housing it is first necessary to remove and disassemble the power steering cylinder as outlined in paragraph 35.

If renewal of the valve body (2—Fig. MF248) is not required, the control valve can be disassembled and overhauled without removal of the unit from the tractor as follows:

Disconnect linkage from control valve by driving the pin (7) from valve rod. Disconnect pressure and return hoses from valve housing, remove hose adapters and apply pressure to valve rod (15) until a small punch can be inserted in adapter hole through hole in valve spool (10). While holding valve spool with punch, unscrew valve rod (15) from spool. Remove valve rod (15), seal (9) and snap ring (8) from valve housing. Remove end cap (1), then push spool and centering spring out seal end of housing.

When reassembling, always renew "O" ring (11) on control valve spool, and carefully insert the assembled valve spool from seal end of body, to avoid damage to "O" ring (11). Re-install snap ring (8), renew seal (9) and install valve end cap (1). Install valve linkage, bleed system as outlined in paragraph 24, and adjust as outlined in paragraph 31.

CONTROL VALVE ADJUSTMENT

Models-F40-MHF50-MF50

31. To adjust the linkage, proceed as follows: With the engine running at 1500 rpm and pin (P—Fig. MF246) removed, loosen jam nut (N) and tighten the adjusting pin (S) until it bottoms. Loosen the jam nuts on the adjusting link and turn sleeve (B) either way as required until pin (P) can be freely installed. Install cotter pins in pin (P) and tighten the linkage jam nuts. Now, back off the adjusting pin (S) seven full turns and tighten jam nut (N).

STEERING CYLINDER

Models TO35-MF35

32. **R&R AND OVERHAUL.** To remove the steering cylinder, first remove the instrument panel assembly

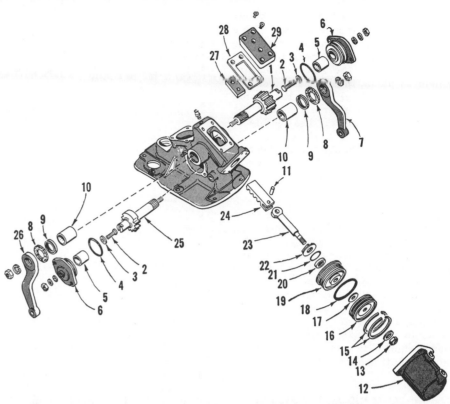

Fig. MF250—Exploded view of TO35 & MF35 power steering cylinder, steering sectors and related parts.

1. Upper sector	8. Dust seal	15. Piston rings	22. Rack stop plate
2. Adjusting screw	9. Seal	16. Piston	23. Piston rod
3. Shims	10. Bushing	17. Thrust washer	24. Power rack
4. "O" ring	11. Coupling pin	18. "O" ring	25. Lower sector shaft
5. Bushing	12. Steering cylinder	19. Cylinder end plate	26. Right pitman arm
6. Side cover	13. Nut	20. Piston rod seal	27. Rack guide
7. Left pitman arm	14. Thrust washer	21. "O" ring	28. Shims

as outlined in paragraph 15A. Disconnect oil lines from cylinder connections, remove the power rack guide cover (29—Fig. MF249) and save the shims located under the cover for reinstallation. Disconnect the pitman arm from the top sector shaft, unbolt

the sector shaft side cover from gear housing and remove the lock nut from adjusting screw (2). Using a screw driver, turn the adjusting screw in and remove the side cover and sector shaft. Remove the four cap screws retaining the power cylinder to the gear

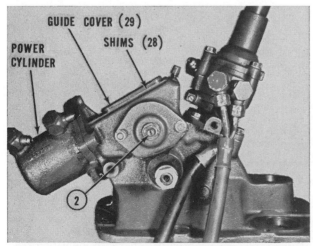

Fig. MF249—TO35 & MF35 power steering system. Sector mesh is adjusted with screws (2) and power rack mesh is adjusted with shims (28).

Fig. MF251—View of TO35 & MF35 power steering unit with the side cover removed. Punch marks on sector gears must be aligned during assembly.

housing and remove the power cylinder and power rack assembly. Using a plastic or lead hammer, bump the cylinder off the cylinder end plate and piston. Remove the self-locking nut retaining piston to piston rod, remove the piston and withdraw the piston rod from the cylinder end plate. The procedure for further disassembly is evident. Refer to Fig. MF250.

When reassembling, renew the piston rings (15), "O" rings and seals. Be sure to tighten the four cylinder retaining cap screws securely. Turn the steering wheel until the punch marked tooth space on the lower sector shaft gear is visible through the left side opening in housing, then install the upper sector shaft so that the punch marked tooth meshes with the marked tooth space on the lower gear. Also, be sure that the upper sector is meshed with the power rack so that rack has full travel in both directions. Refer to Fig. MF251. Install the upper sector shaft side cover and adjust the unit as outlined in paragraph 38.

Models MHF202-MF202-MF204

33. An installation view of the power steering cylinder is shown in Fig. MF252. The only overhaul work which can be performed is the renewal of seals. The procedure for renewing the seals is evident after an examination of the unit. If cylinder parts other than seals are damaged, it is recommended that complete cylinder be renewed.

Models F40-MH50-MF50
Before Ser. No. 516788

34. **R&R AND OVERHAUL.** To remove the power steering cylinder, remove the grille screens, remove the cap screws retaining the valve assembly (Fig. MF246) to top of cylinder and move the valve unit out of way. Bump the pin (X) down to release the cylinder rod end and remove the cap screws (CS). Swing the cylinder outward and as far toward left as pos-

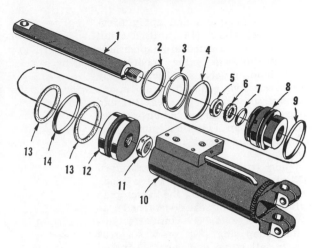

Fig. MF253 — Exploded view of the power steering cylinder used on models F40, MH50 and early MF50.

1. Piston rod
2. Retainer ring
3. Spacer
4. Retainer ring
5. Wiper seal
6. Leather washer
7. "O" ring
8. Tube head
9. "O" ring
10. Cylinder
11. Nut
12. Piston
13. Back-up washer
14. "O" ring

sible. Remove pin (Y) and withdraw the cylinder.

Remove ring (2—Fig. MF253) and spacer (3). Push the tube head (8) inward and remove retainer ring (4). Withdraw the rod, piston and head assembly. Remove tube head from rod. When reassembling, renew all "O" rings and seals and be sure retainer rings are properly seated in their grooves. Note: Removal of piston (12) from rod (1) is not recommended.

Before installing the cylinder, remove the rod end bracket (Fig. MF-246), extract pin (X), then install the bracket and cylinder. Bleed the power steering system as in paragraph 24.

MF50 After Ser. No. 516787

35. **R&R AND OVERHAUL.** To remove the power steering cylinder, remove grille screens and disconnect pressure and return lines and control linkage from control valve. Refer to Fig. MF246 and bump pin (X) to release rod end of piston, remove the two cap screws (CS). Swing the cylinder and control valve forward as far as possible, remove pin (Y) and lift the cylinder off the tractor.

To disassemble the cylinder, refer to Fig. MF248, remove the four tie-

bolts (18) and remove end cap (26) and cylinder tube (25). Remove any dirt, paint or burrs from exposed end of piston rod (19) and withdraw rod and piston assembly from valve housing (2). Unless necessary for the renewal of parts, it is not recommended that piston (21) or nut (22) be removed from piston rod. Inspect piston rod (19), piston (21) and cylinder tube (25) for wear or scoring, renew all seals and "O" rings and reassemble by reversing the disassembly procedure. The control valve may be overhauled while the cylinder is removed, as outlined in paragraph 30.

STEERING PEDESTAL
Models F40-MH50-MF50

36. **R&R AND OVERHAUL.** The power steering pedestal is shown installed in Fig. MF246. To remove the pedestal, first remove the grille screens and the left side panel. Disconnect drag link from pedestal arm and remove pin (P).

Bump pin (X) downward to release the cylinder rod end and remove the cap screws (CS). Swing the cylinder and valve assembly outward and as far toward left as possible. Remove pin (Y) and lay cylinder and valve assembly out of way. Loosen the cap screw retaining the steering arm to lower end of pedestal shaft, unbolt pedestal from the front support casting and lift the pedestal assembly from tractor.

NOTE: There are no blind splines to facilitate correct assembly of the cylinder connecting arm (22—Fig. MF-254) with respect to pedestal shaft (16). It is important, therefore, that the relative position of the two parts be marked or otherwise identified before disassembly. Remove snap ring (15) and press the pedestal shaft (16)

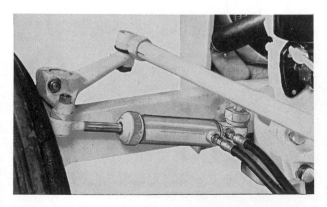

Fig. MF252—Power steering cylinder installation on left side of models MHF202 MF202 and MF204. The right cylinder installation is similar.

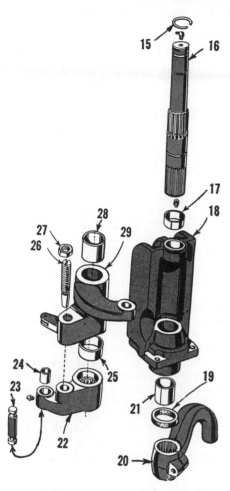

Fig. MF254 — Power steering pedestal and components used on models F40, MH50 and MF50.

15. Snap ring	23. Pin
16. Pedestal shaft	24. Bushing
17. Bushing	25. Bushing
18. Pedestal	26. Adjusting screw
19. Dust seal	27. Nut
20. Steering arm	28. Bushing
21. Bushing	29. Valve connecting
22. Cylinder connect-	arm
ing arm	

down and out of pedestal. Be sure to renew any parts which show an excessive amount of wear. All bushings are pre-sized and available for service installation.

After pedestal is installed on tractor, adjust the linkage as outlined in paragraph 31.

GEAR UNIT
Models F40-MH50-MF50

37. The gear unit used on models with power steering is the same as the unit used on models without power steering. The procedure for adjusting the unit is outlined in paragraphs 17, 18 and 19. Removal, reinstallation and overhaul of the unit is outlined in paragraphs 20, 21 and 22.

Models TO35-MF35

38. **ADJUSTMENT.** Before attempting to adjust the steering gear, first make certain that the gear housing is properly filled with lubricant, then disconnect the drag links from pitman arms to remove load from the gear unit. As shown in Fig. MF242, the wormshaft is carried in needle type bearings which require no adjustment. The steering valve thrust bearings are adjusted by tightening the bearing nut to a torque of 20-30 Ft.-Lbs., but this should be done only when the valve unit is being serviced. To adjust the sector gear and rack mesh, proceed as follows:

Loosen both sector shaft adjusting screws (2—Fig. MF250) at least four turns, remove the rack guide cover (29) and turn the steering wheel to the mid (or straight ahead) position. Using a screw driver, turn the adjusting screw (2) for the lower sector shaft until a spring scale pull of 1¼-1¾ pounds (measured at rim of steering wheel) is required to turn the wheel through a 3-inch arc. Tighten the adjusting screw lock nut.

With the lower sector adjusted, turn the adjusting screw (2) for the upper sector shaft until a spring scale pull of 2-2½ pounds is required when measured as before. Tighten the lock nut.

Now, install the rack guide cover (29) and vary the number of shims (28), located under the cover, until a spring scale pull of 2⅛-2¾ pounds is required to turn the steering wheel through a 3-inch arc at mid-point of the turning range.

39. **OVERHAUL.** The steering gear can be overhauled without removing the gear housing from tractor, as follows:

Remove the instrument panel assembly as outlined in paragraph 15A and the steering valves as outlined in paragraph 28. Unbolt and remove the power rack guide cover (29—Fig. MF-250) from top of housing and save shims (28) for reinstallation. Disconnect both pitman arms from the sector shafts. Unbolt the upper sector shaft side cover (6) from housing and remove lock nut from adjusting screw (2). Using a screw driver, turn the adjusting screw in and remove the side cover and sector shaft. Remove the other side cover and sector shaft in the same manner. Remove the four cap screws retaining the power cylinder to the gear housing and remove the power cylinder and rack assembly. Withdraw the wormshaft and ball nut assembly from housing. Do not disassemble the ball nut assembly as component replacement parts are not available. If shaft and/or ball nut are damaged, renew the complete as-

sembly. The need and procedure for further disassembly and/or overhaul is evident.

Shims (3) on the adjusting screws (2) are available in thicknesses of 0.063, 0.065, 0.067 and 0.069. When reassembling, use the proper combination of shims to provide a very slight amount of clearance between adjusting screw head and slot in sector shafts.

When reassembling, install the steering wormshaft and ball nut assembly; then install the steering valves as outlined in paragraph 28. Install the power cylinder and rack assembly and tighten the retaining cap screws securely. Turn the steering shaft until the ball nut raises to a position where the lower gear tooth on the ball nut is clearly visible through the right side opening in housing. Hold the lower sector shaft and side cover assembly with blank portion of gear down and install shaft through right side opening in housing. Mesh sector gear with ball nut so that lower tooth on ball nut meshes with the groove next to the blank portion of sector shaft gear.

Turn the steering wheel until the punch marked tooth space on the lower sector shaft gear is visible through the left side opening in housing, then install the upper sector shaft so that the punch marked tooth meshes with the marked tooth space on the lower gear. Also, be sure that the upper sector is meshed with the power rack so that rack has full travel in both directions. Refer to Fig. MF251. When assembly is complete, adjust the unit as outlined in paragraph 38.

Models MHF202-MF202-MF204

40. **ADJUSTMENT.** Before attempting to adjust the steering gear, first make certain that the gear housing is properly filled with lubricant, then disconnect the drag links from pitman arms to remove load from the gear unit. As shown in Fig. MF245, the wormshaft is carried in needle type bearings which require no adjustment. The steering valve thrust bearings are adjusted by tightening the bearing nut to a torque of 20-30 Ft.-Lbs., but this should be done only when the valve unit is being serviced. To adjust the sector gear backlash, proceed as follows:

Loosen both sector shaft adjusting screws at least four turns and turn the steering wheel to the mid (or straight ahead) position. Using a screw driver, turn the adjusting screw for the lower sector shaft until a spring scale pull of 1¼-1¾ pounds (measured at rim of steering wheel) is required to turn the wheel through a 3-inch arc. Tighten the adjusting screw lock nut.

With the lower sector adjusted, turn the adjusting screw for the upper sector shaft until a spring scale pull of 2-2½ pounds is required when measured as before. Tighten the lock nut.

41. **OVERHAUL.** The steering gear can be overhauled without removing the gear housing from tractor by following the general procedure for the power steering unit in paragraph 39. Keep in mind, however, that the gear unit is not equipped with the attached power cylinder as used on the TO35 and MF35.

When assembly is complete, adjust the unit as outlined in paragraph 40.

NON-DIESEL ENGINE AND COMPONENTS

R&R ENGINE WITH CLUTCH

42. To remove the engine and clutch, first drain cooling system and if engine is to be disassembled, drain oil pan. Disconnect head light wires, remove hood and on models so equipped, disconnect radius rods and/or drag links. Disconnect radiator hoses. On model MF204 drain "Revers-O-Matic" unit and power steering, remove front grille closure and disconnect the two lines leading to oil cooler radiator. Remove the two lines leading to the "Revers-O-Matic" oil filter and unbolt and remove filter. Remove starter then disconnect and remove the long line to the oil cooler. Disconnect the pressure and return lines at power steering valve and the four cylinder lines at the two junction blocks under battery platform. Support tractor under transmission housing and unbolt front support casting from engine. Roll front axle, support and radiator as an assembly away from tractor. Shut off fuel and remove fuel tank. Disconnect the heat indicator sending unit from water outlet elbow and cable from starting motor. Disconnect wires from coil and generator, tractormeter cable from generator, choke rod from carburetor and oil gage line from right side of cylinder block. Remove the air cleaner pipe, loosen the U-bolt assembly from front end of throttle rod and disconnect exhaust pipe from manifold. Disconnect battery cables and remove battery. Unbolt battery platform from engine, support engine in a hoist and unbolt engine from transmission case.

CYLINDER HEAD

43. **REMOVE AND REINSTALL.** To remove the cylinder head, drain cooling system, shut off fuel and remove fuel tank. Disconnect governor-to-carburetor rod and loosen U-bolt assembly from front end of throttle rod. Disconnect choke rod from carburetor, loosen carburetor to air cleaner

pipe hose and unbolt exhaust pipe from manifold. Unbolt manifold from cylinder head and remove manifold and carburetor assembly. Remove upper radiator hose, rocker arm cover and rocker arms assembly. Unbolt and remove cylinder head from tractor.

When reinstalling the cylinder head, tighten the stud nuts from the center outward and to a torque value of 70-75 Ft.-Lbs. Valve tappet gap is given in paragraph 44.

VALVES AND SEATS

44. Valve tappet gap is 0.015 cold or 0.013 hot. All valves are equipped with stem caps.

Intake valves seat directly in cylinder head and the valve stems are equipped with neoprene oil guards. Exhaust valves, except on MF50 LP-Gas, are equipped with positive type rotators. Exhaust valves on all models are equipped with renewable type seat inserts. Replacement exhaust valve seat inserts are 0.010 oversize. Desired interference fit of new inserts is 0.003-0.005.

Intake valves have a face and seat angle of 30 degrees with a desired seat width of 5/64 inch. Exhaust valves have a face angle of 44 degrees and a seat angle of 45 degrees. Desired exhaust valve seat width is 5/64 inch. Seats can be narrowed, using 15 and 75 degree stones.

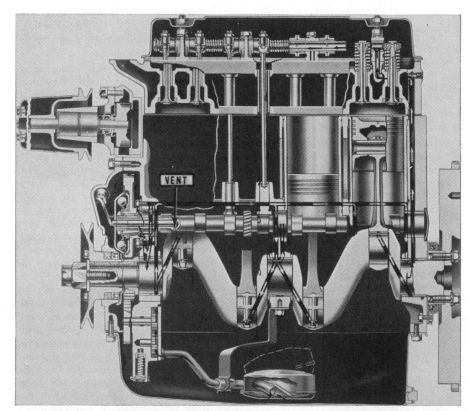

Fig. MF255—Sectional view 134 cubic inch engine. Bore and stroke are 3 5/16 inches and 3⅞ inches respectively. Crankshaft end play is controlled by the flanged center main bearing.

Valve stem diameter is 0.3141-0.3149 for the intake, 0.3124-0.3132 for the exhaust.

VALVE GUIDES

45. The pre-sized intake and exhaust valve guides are interchangeable and can be driven from cylinder head if renewal is required. Guides should be pressed into the cylinder head, using a piloted drift 0.002 smaller than bore of guide, until port end of guide is 1 11/16 inches from gasket surface of cylinder head.

Desired valve stem clearance in guides is 0.0016-0.0034 for the intake, 0.0033-0.0051 for the exhaust.

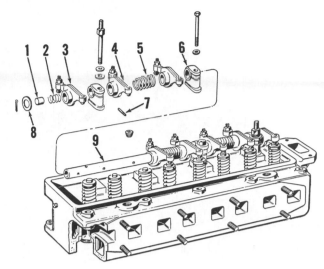

Fig. MF257—Partially exploded view of rocker arms and shaft assembly typical of all except TO-35 & MF35 Diesel. Maximum allowable clearance between the rocker arms and shaft is 0.006.

1. Cork plug
2. Spring, short
3. Rocker arm, right hand
4. Rocker arm, left hand
5. Spring, long
6. Shaft support
7. Support pin
8. Washer
9. Rocker arm shaft

VALVE SPRINGS

46. Intake and exhaust valve springs are interchangeable. Renew any spring which is rusted, discolored or does not meet the test specifications which follow:

Free length . . . . 2 1/16 inches
Lbs. test @ 1 45/64 inches 47-53
Lbs. test @ 1 27/64 inches 96-104

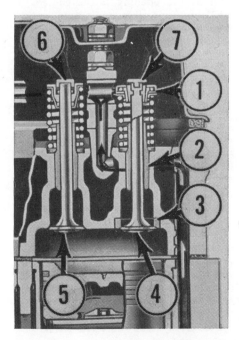

Fig. MF256—Valve arrangement on gasoline models. LP-Gas models are similar, except valve rotators are not used. Both the intake (5) and exhaust (4) valves are equipped with stem caps (6 and 7). Exhaust valves are equipped with seat inserts (3) and positive type rotators (1). Normal servicing of faulty rotators consists of renewing the units.

VALVE ROTATORS
All Gasoline Models

46A. Normal servicing of the positive type exhaust valve rotators ("Rotocaps") consists of renewing the units. It is important, however, to observe the valve action after engine is started. The valve rotator action can be considered satisfactory if the valve rotates a slight amount each time the valve opens. A cutaway view of a typical "Rotocap" installation is shown in Fig. MF256.

VALVE TAPPETS

47. Mushroom type tappets (cam followers) operate directly in machined bores of the cylinder block. The 0.5615-0.5620 diameter tappets are furnished in standard size only and should have a diametral clearance of 0.0005-0.0025 in the block bores. Tappets can be removed after removing camshaft as outlined in paragraph 52.

Refer to paragraph 44 for valve tappet gap.

ROCKER ARMS

48. Rocker arms and shaft assembly can be removed after removing fuel tank and rocker arm cover. The rocker arms, being right and left hand assemblies, are not interchangeable. Desired clearance between new rocker arms and new rocker arm shaft is 0.001-0.003. Renew shaft and/or rocker arms if diametral clearance exceeds 0.006. Diameter of new rocker arm shaft is 0.622-0.623. Oil holes in rocker arm shaft face toward valve spring. When installing corks in end of shaft, push them in only far enough to insert the cotter pins.

Refer to paragraph 44 for tappet gap.

VALVE TIMING

49. Valves are properly timed when single punch marked tooth on crankshaft gear is meshed with double punch marked tooth space on camshaft gear.

To check valve timing when engine is assembled, first adjust the number one cylinder intake valve tappet gap to 0.020. Insert a 0.005 feeler gage and crank engine over slowly until a slight resistance occurs when trying to withdraw the feeler gage from between valve stem and rocker arm. At this time, the "DC" mark on flywheel should be in register with groove in timing hole on left side of engine.

Reset valve to the proper tappet gap given in paragraph 44.

TIMING GEAR COVER

50. To remove the timing gear cover, drain cooling system, disconnect head light wires and remove hood assembly. Disconnect radiator hoses and on models so equipped, disconnect the radius rods and/or drag link. Support tractor under transmission housing, unbolt the front axle support from engine and roll the front axle, support and radiator as an assembly away from tractor. Remove fan blades and disconnect spring and rod from governor lever. Remove starting jaw and crankshaft pulley. Remove the three cap screws retaining oil pan to timing gear cover and loosen the remaining oil pan cap screws. Unbolt and remove timing gear cover from engine.

The crankshaft front oil seal (29—Fig. MF258 can be renewed at this time and should be installed with lip facing the engine.

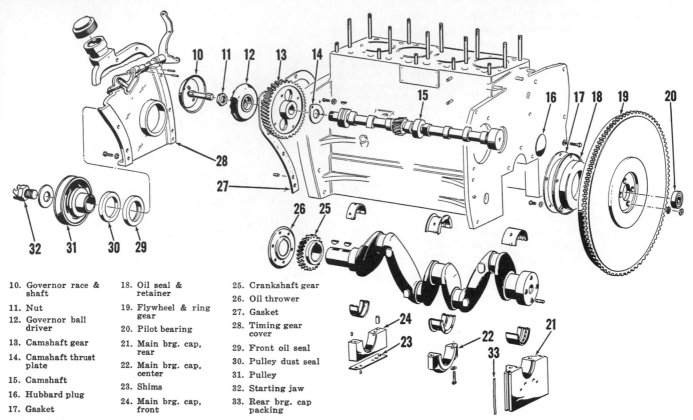

10. Governor race & shaft	18. Oil seal & retainer	25. Crankshaft gear
11. Nut	19. Flywheel & ring gear	26. Oil thrower
12. Governor ball driver	20. Pilot bearing	27. Gasket
13. Camshaft gear	21. Main brg. cap, rear	28. Timing gear cover
14. Camshaft thrust plate	22. Main brg. cap, center	29. Front oil seal
15. Camshaft	23. Shims	30. Pulley dust seal
16. Hubbard plug	24. Main brg. cap, front	31. Pulley
17. Gasket		32. Starting jaw
		33. Rear brg. cap packing

Fig. MF258—Engine camshaft, crankshaft and timing gears typical of all except TO35 & MF35 Diesel. Crankshaft end play is controlled by the flanged center bearing.

TIMING GEARS

51. Timing gears can be renewed after removing timing gear cover as outlined in paragraph 50. Remove governor race and shaft assembly (10 —Fig. MF258) and the nut retaining camshaft gear to camshaft. Remove governor ball driver assembly (12) and using a suitable puller, remove the camshaft timing gear, being careful not to damage the governor shaft bore in camshaft. The procedure for removing the crankshaft gear is evident.

Recommended timing gear backlash is 0.002. Gears are available in oversizes and undersizes of 0.001 and 0.002 to facilitate obtaining the desired backlash. Gear markings are as follows: Standard, "S"; Undersize, "U"; Oversize, "O".

NOTE: For production reference, a timing gear backlash number is stamped on block front face. When renewing timing gears, any combination of oversize and undersize gears in which the total equals the stamped reference number should give the correct backlash.

During installation, mesh the single punch marked tooth on crankshaft gear with the double punch marked tooth space on camshaft gear and when camshaft gear is being drifted

on, it is advisable to remove oil pan and buck up the shaft with a heavy bar.

Tighten the camshaft gear retaining nut to a torque of 65-70 Ft.-Lbs.

CAMSHAFT

52. To remove the camshaft, first remove the camshaft timing gear as outlined in paragraph 51. Remove fuel tank, rocker arm cover, rocker arms and shaft assembly and push rods. Remove the ignition distributor and oil pan and block-up or support tappets (cam followers). Remove the camshaft thrust plate (14—Fig. MF-258) and withdraw camshaft from front of engine.

Camshaft journals ride directly in three machined bores in cylinder block with a normal diametral clearance of 0.0025-0.0045. If clearance exceeds 0.007, renew camshaft and/or cylinder block or make up and install bushings.

Camshaft journal diameters are as follows:

Front 1.8090-1.8095
Center 1.7460-1.7465
Rear 1.6835-1.6840

Before installing the camshaft, check the vent opening directly behind the number one cylinder exhaust

cam. If the opening is obstructed, it will be impossible to obtain satisfactory operation of the governor.

Desired camshaft end play of 0.003-0.007 is controlled by thrust plate (14). Renew the plate if end play exceeds 0.008.

ROD AND PISTON UNITS

53. Connecting rod and piston units are removed from above after removing cylinder head and oil pan. Pistons and rods are installed with the rod correlation marks facing the camshaft. Replacement rods are not marked and should be installed with the oil spray hole (lower end of rod) facing away from camshaft side of engine. Tops of pistons are installed with arrow marked front facing toward front of engine.

Connecting rod bolt torque is 35-40 Ft.-Lbs.

PISTONS, SLEEVES AND RINGS

54. Aluminum alloy, cam ground pistons are supplied only in the standard size and are available only in a sleeve and piston kit consisting of pistons, pins, rings and sleeves. Recommended piston skirt clearance is .0012. Piston clearance is checked with a spring scale pull of 5-10 lbs., using a 0.002 x ½ inch feeler gage. Wear limit

of worn pistons and sleeves is when a 0.005 x ½ inch feeler gage requires less than 5 lbs. pull on the spring scale to withdraw it.

With the piston and connecting rod assembly removed from the cylinder block, use a suitable puller to remove the sleeve. Before installing the wet type sleeve, clean all cylinder block sealing surfaces. The top of the sleeve should extend 0.002-0.0045 above the top surface of the cylinder block. If this standout is in excess of 0.0045, check for foreign material under sleeve flange. Excessive standout will cause water leakage at cylinder head gasket. To facilitate installation of sleeves, use a lubricant (palm oil or vaseline) on the two neoprene sealing rings.

There are three ⅛ inch wide compression rings and one 3/16 inch wide oil control ring per piston. Recommended end gap is 0.010-0.018 for the oil ring, 0.010-0.020 for the compression rings. Recommended side clearance for all compression rings is 0.0035-0.005; for the oil control ring, 0.002-0.0035.

Standard size cylinder bore is 3.313-3.315.

PISTON PINS

55. The 0.8591-0.8593 diameter floating type piston pins are retained in piston bosses by snap rings and are available in standard and 0.003 and 0.005 over-size. The bushing in the upper end of connecting rod has a very thin wall and if sized by reaming, do so only with a fluted type, taking very light cuts. Be sure oil hole in bushing registers with oil hole in top end of the connecting rod and clean same thoroughly after sizing the bushing. Pin should be fitted to a 0.0002-0.0006 clearance in the rod and a minus .0001 to plus .0003 clearance in the piston.

CONNECTING RODS AND BEARINGS

56. Connecting rod bearings are of the shimless, non-adjustable, slip-in precision type renewable from below after removing oil pan. When installing new bearing shells, be sure that the projection engages milled slot in rod and cap and the rod and rod cap correlation marks are in register. Replacement rods are not marked and should be installed with the oil spray hole facing away from camshaft side of engine. Bearings are available in 0.002, 0.010 and 0.020 undersize, as well as standard.

Crankpin diameter 1.9365-1.9375
Diametral clearance 0.0015-0.0025
Renew if clearance exceeds .. 0.0045
Side clearance 0.005-0.011
Renew if side clearance exceeds 0.014
Rod length C to C...... 6.373-6.377
Cap screw torque ...... 35-40 ft. lbs.

CRANKSHAFT AND BEARINGS

57. Crankshaft is supported on 3 shimless, non-adjustable, slip-in, precision type main bearings, renewable from below without removing the crankshaft. The rear main bearing cap contains packing (33—Fig. MF258) on each side of the filler block to prevent oil leakage into the clutch housing. This packing is in addition to the separate crankshaft rear seal (18). To remove the rear main bearing cap and filler block, first remove the two cap screws which retain the crankshaft rear oil seal retainer to the bearing cap and then remove two bearing cap retaining screws.

Bearing inserts are available in standard, 0.002, 0.010 and 0.020 undersizes.

Normal crankshaft end play of 0.002-0.006 is controlled by the flanged center main bearing.

To remove crankshaft it is necessary to remove engine, clutch, flywheel, rear oil seal, timing gear cover, oil pan, and bearing caps.

Check the crankshaft journals for wear, scoring and out-of-round condition against the values listed below:
Journal diameter 2.249-2.250
Diametral clearance 0.0015-0.0025
Cap screw torque........85-95 ft.-lbs.

CRANKSHAFT REAR OIL SEAL

58. Crankshaft rear oil seal (18—Fig. MF258) is contained in a one piece retainer and serviced only as an assembly. To renew the seal, first separate the engine from the transmission case as in paragraph 174 or 190, and remove the flywheel and oil pan. Remove the three seal retainer to crankcase cap screws and the two seal retainer to rear main bearing retaining cap screws.

FLYWHEEL

59. To remove flywheel, separate tractor as outlined in paragraph 174 or 190, and remove clutch unit (or torque converter) from the flywheel. The starter ring gear can be renewed after removing the flywheel. To install a new ring gear, heat same to 500 deg. F. and install on flywheel with beveled end of teeth facing timing gear end of engine. One flywheel mounting bolt hole is off-center.

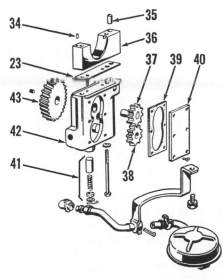

Fig. MF259—Engine oil pump on all except TO35 & MF35 Diesel is mounted on bottom side of the crankshaft front main bearing cap. Shims (23) control backlash between oil pump gear and crankshaft gear.

35. Dowel pin
36. Main bearing cap, front
37. Driven gear and shaft
38. Idler gear
39. Gasket
40. Cover
41. Oil pressure relief valve
42. Pump body
43. Pump drive gear

OIL PUMP

60. Gear type pump, shown in Fig. MF259, is bolted to bottom of number one main bearing cap and is gear driven from crankshaft pinion. Pump is accessible after oil pan is removed. Shims (23) interposed between pump body and main bearing cap are varied to obtain the desired 0.005-0.010 backlash between the crankshaft pinion and the aluminum gear on the pump shaft.

Check the pump internal gears for backlash, which should not exceed 0.007. The recommended diametral clearance between the gears and pump body is 0.003-0.004. Gear side clearance is 0.004-0.008. A lead gasket, 0.007 thick, is placed between pump body and cover to control side clearance (end play). Presized pump shaft bushings are available for service and have a running clearance of 0.0035-0.0065.

RELIEF VALVE

61. Plunger type relief valve (41—Fig. MF259) is located in oil pump body and can be adjusted with spacers inserted under the spring so as to maintain a pressure of 20-30 psi at 2200 engine rpm. At idle speed, pressure should be not lower than 15 psi. The plunger type valve should fit in its bore with a 0.003-0.005 clearance. The relief valve spring should have a free length of 2 inches and should check 7¾-8¼ pounds at a working length of 1⅜ inches.

STANDARD DIESEL ENGINE AND COMPONENTS

R&R ENGINE WITH CLUTCH

62. To remove the engine and clutch as a unit, first drain cooling system and if engine is to be disassembled, drain oil pan. Shut off the fuel. Remove the hood, disconnect injection nozzle leak-off line and both lines from thermo-start reservoir. Disconnect line from water trap and sediment bowl.

Remove two rear fuel tank support cap screws. Remove nuts and springs on two front tank support bolts and remove tank.

Disconnect upper and lower radiator hoses at radiator. Disconnect both drag links and radius rods at the forward end. Disconnect radiator support rod at thermostat housing. Remove air cleaner to manifold hose. Block tractor up under transmission housing, unbolt axle support from engine and roll the axle support and wheels assembly away from tractor.

Disconnect thermo-start aid wire, battery cable and wiring harness from starter solenoid and generator. Disconnect heat indicator sending unit from thermostat housing. Unbolt exhaust pipe from manifold. Disconnect fuel line from transfer pump, throttle rod and stop rod from the injection pump, two fuel lines from second stage fuel filter to transfer pump and the tachourmeter cable from injection pump. Disconnect oil pressure banjo fitting on left side of engine. Remove the starting motor.

Swing engine in a hoist (or support by other means), unbolt the engine from the transmission housing and separate the tractor halves.

CYLINDER HEAD

63. **REMOVE AND REINSTALL.** To remove the cylinder head, drain cooling system, shut off fuel and remove fuel tank. Loosen the rocker arm cover breather tube clamp, then remove rocker arm cover. Disconnect fuel lines from injection nozzles.

Remove rocker arm assembly and push rods. Note: Valve caps are of different types. The exhaust valves have free type rotator caps while the intake valves have wear caps only (smaller than exhaust).

Remove air cleaner to manifold hose. Disconnect thermo-start aid wire. Remove rocker arm cover breather pipe. Unclip harness and heat indicator sending wire from right side of block. Disconnect exhaust pipe from manifold.

Remove capscrews retaining thermostat housing to cylinder head, then unbolt and remove the cylinder head.

Note: Copper side of gasket faces block. Gasket is also marked TOP, on the upper face. Refer to paragraph 141 for information regarding the precombustion chambers.

Assembly is the reverse of removal. When reinstalling the cylinder head, tighten the head retaining nuts and capscrews from the center outward and to a torque of 100-105 ft.-lbs. Refer to paragraph 64 for valve tappet gap.

VALVES AND SEATS

64. Valve tappet gap (cold) is 0.012 for the inlet and 0.008 for the exhaust. The intake valves are equipped with wear caps, while the exhaust valves have free type rotators. Valves are numbered in sequence from front to back; with Nos. 1 & 2 in No. 1 cylinder, Nos. 3 & 4 in No. 2 cylinder, etc.

The intake valves on production engines seat directly in cylinder head as shown in view A in Fig. MF261. Valve face angle is 45 degrees and desired seat angle in head is 45 degrees. Intake valve seats can be refaced so long as a perceptible step remains. If, however, reseating will remove the step, it will be necessary to counterbore the head and install a seat insert as shown in view B. Seat insert should have an interference fit in head and should be located 0.040 below face of head. Seat insert can be refaced so long as metal is not removed from head at point (Y).

Exhaust valves seat on renewable seat inserts as shown in view (C).

Fig. MF260—Three-quarter section view of Standard Diesel engine. The unit is fitted with a C.A.V. distributor type injection pump and Pintaux type injector nozzles. Cylinder sleeves are of the slip-in, dry type and main and rod bearings are not adjustable.

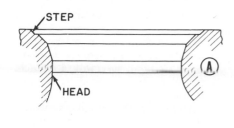

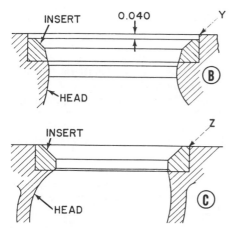

Fig. MF261—Sectional views showing the intake and exhaust valve seats on the Standard Diesel. Refer to text.

Valve face angle is 45 degrees and desired seat angle is 45 degrees. Seat insert can be refaced so long as metal is not removed from head at point (Z).

NOTE: Where the intake and exhaust valves as well as their seats have been recut to their maximum limits, the valves will sink too deeply into the head. The distance between head of valve and face of cylinder head must never exceed 0.100.

Valve stem diameter is 0.3107-0.3112 for the intake, 0.3727-0.3732 for the exhaust.

VALVE GUIDES

65. The pre-sized intake and exhaust valve guides are not interchangeable and can be driven from cylinder head if renewal is required. Guides should be pressed into the cylinder head, using a piloted drift 0.002 smaller than bore of guide, until end of guide is ⅜-inch above the spring seat for the intake and $\frac{9}{16}$-inch for the exhaust.

Desired valve stem diametral clearance in the guides is 0.0008-0.0023 for the intake, 0.0013-0.0028 for the exhaust.

VALVE SPRINGS

66. Intake and exhaust valve springs are interchangeable. Renew any spring which is rusted, discolored or does not meet the test specifications which follow:

Inner spring:
    Free length....1.6 inches (approx.)
    Lbs. test @ 1.22 inches........ 22
    Lbs. test @ ⅞ inch............41.9
Outer spring:
    Free length....1.7 inches (approx.)
    Lbs. test @ 1.316 inches....... 40
    Lbs. test at 31/32 inch........ 77

VALVE ROTATORS

67. The model TO35 Diesel utilizes free type rotator caps (19—Fig. MF-262) on the exhaust valves.

VALVE TAPPETS

68. The TO35 Diesel valve tappets (cam followers) are of the mushroom type and operate directly in machined bores of the cylinder block. The 0.5618-0.5621 diameter tappets are furnished in standard size only and should have a diametral clearance of 0.0002-0.0012 in the block bores. Tappet bore in block is 0.5623-0.5630. Tappets can be removed after removing camshaft as outlined in paragraph 73. Refer to paragraph 64 for valve tappet gap.

ROCKER ARMS

69. Rocker arms and shaft assembly can be removed after removing the fuel tank and rocker arm cover. The rocker arms, being right and left hand assemblies, are not interchangeable. Each rocker arm is fitted with a bushing which can be renewed if rocker arm is in otherwise good condition. Ream bushing after installation, if necessary, to provide a free fit for the shaft. Replacement arms contain a factory installed bushing.

Refer to paragraph 64 for valve tappet gap.

TIMING SPROCKET COVER

70. To remove the timing chain and sprocket cover, drain cooling system, disconnect headlight wires and remove hood assembly. Disconnect radiator hoses and radiator support rod at thermostat housing. Disconnect drag links and radius rods. Remove air cleaner to manifold hose. Support tractor under transmission housing, unbolt the front axle support from engine and roll the front axle, support and radiator away from tractor.

Refer to Fig. MF263. Remove fan assembly, loosen generator strap and remove fan belt. Bend lock washer lip away from starter jaw and remove the starting jaw. Remove fan belt pulley. Remove capscrews retaining cover to block and remove cover.

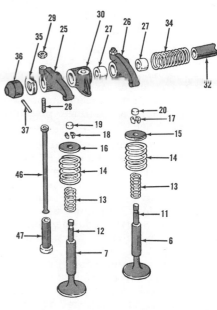

Fig. MF262—Valves, rocker arms, shaft and associated parts on Standard Diesel.

6. Intake valve guide	26. Rocker arm, intake
7. Exhaust valve guide	27. Rocker arm bushings
11. Intake valve	28. Tappet adjusting screw
12. Exhaust valve	29. Lock nut
13. Inner spring	30. Rocker shaft pedestal
14. Outer spring	32. Rocker arm shaft
15. Intake valve spring retainer	34. Spring, intermediate
16. Exhaust valve spring retainer	35. Spring, outer
17. Intake valve spring keepers	36. Spring retaining collar
18. Exhaust valve spring keepers	37. Collar pin
19. Rotator cap	46. Push rod
20. Wear cap	47. Tappet
25. Rocker arm, exhaust	

Crankshaft front oil seal, which is located in the cover, can be renewed at this time if desired.

Reassembly is the reverse of removal procedure. Be sure to properly locate timing sprocket cover on its dowels.

Fig. MF263—Front view of Standard Diesel engine with front axle assembly removed.

25

TIMING SPROCKETS, CHAIN AND TENSIONER

71. The timing sprockets, timing chain and chain tensioner can be renewed after removing the timing sprocket cover as outlined in paragraph 70 and the rocker arms and shaft assembly as outlined in paragraph 69. Then proceed as follows:

Remove Woodruff key used to position fan belt pulley and slide oil pump drive gear (Fig. MF264) from crankshaft; at this same time, remove a second Woodruff key. Remove four capscrews retaining oil pump assembly to front main bearing housing and remove oil pump. Remove two capscrews retaining chain tensioner to front bearing housing and remove tensioner. Remove capscrews retaining camshaft sprocket (Fig. MF265) to camshaft gear hub and slide sprocket from hub; at this time, timing chain can be removed. Slide crankshaft sprocket from crankshaft, but be sure to save all shims between sprocket and crankshaft for later installation. Remove the third Woodruff key from crankshaft.

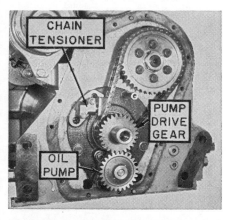

Fig. MF264—Standard Diesel timing chain and sprockets. Chain lubrication is provided by an oil passage in the chain tensioner.

Fig. MF265 — Timing chain and sprockets can be removed after removing oil pump and oil pump drive gear on Standard Diesel.

To reassemble, install the same number of shims previously removed from between the crankshaft sprocket and crankshaft. Then turn the crankshaft Woodruff keyways to a 9 o'clock position as shown in Fig. MF266. Install a Woodruff key and the crankshaft sprocket. IMPORTANT: Do **not** move the crankshaft after sprocket installation until mentioned subsequently.

Install the rocker arms and shaft assembly. Now, turn camshaft clockwise until the valves of No. 4 cylinder are fully closed. Adjust the number 4 cylinder tappet gap to 0.025 for the intake, 0.023 for the exhaust. (This adjustment is for timing purposes only). Continue turning the camshaft clockwise until both valves are in a "rocking" position (exhaust valve just about to close and intake valve just about to open). IMPORTANT: Do **not** move the camshaft beyond this position.

At this time, turn the crankshaft counterclockwise until the keyways are at a 6 o'clock position (straight down). Now install the camshaft sprocket, timing chain and sprocket retaining capscrews. Camshaft sprocket can be installed in one position only, due to off-center capscrew holes. Install the chain tightener and retaining capscrews, as follows:

With the lock screw (13—Fig. MF-267) and tab (12) removed, insert a No. 4 Allen wrench into the housing (1) through the screw hole. Then install the adjusting cylinder (11), spring (7) and inner cylinder (4) into the housing bore and on to the Allen wrench. Rotate Allen wrench counterclockwise to hold assembly in place.

Install chain tightener on front main bearing housing with spacer (3) interposed between tightener and housing. Install lock washers (10) and retaining capscrews (9).

After unit is installed on front main bearing housing, adjust the chain tightener by turning the Allen wrench clockwise sufficiently to take up timing chain slack. Install tab washer (12) and screw (13).

Install Woodruff key, oil pump assembly and retaining screws. Install the oil pump drive gear on crankshaft; then install the fan pulley Woodruff key.

Remainder of installation procedure is evident. Be sure to correctly set valve tappet gap as outlined in paragraph 64.

INJECTION PUMP DRIVE

72. The injection pump is driven from a gear machined on the camshaft, through an idler gear located in injection pump drive adapter housing as shown in Fig. MF268. To remove the injection pump drive housing, first remove the injection pump as outlined in paragraph 137, disconnect the tractormeter drive adapter from front of housing and unbolt and remove housing (2), together with idler

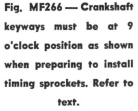

Fig. MF266 — Crankshaft keyways must be at 9 o'clock position as shown when preparing to install timing sprockets. Refer to text.

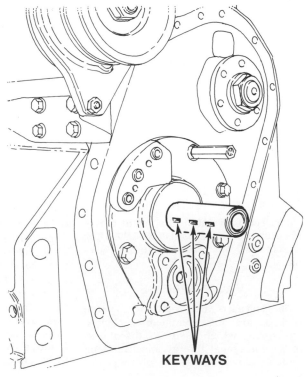

KEYWAYS

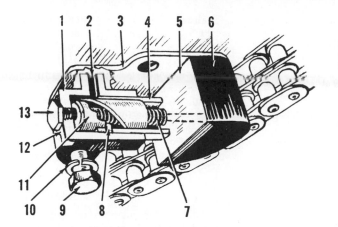

Fig. MF267 — Sectional view of timing chain tensioner used on Standard Diesel.

1. Housing
2. Oil passage
3. Spacer
4. Inner cylinder
5. Pressure plate
6. Composition block
7. Compression spring
8. Lock pin
9. Cap screw
10. Lock washer
11. Adjusting cylinder
12. Tab lock
13. Lock screw

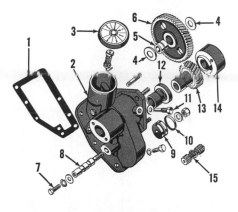

Fig. MF268 — Exploded view of injection pump adapter housing and gears used on 23C Standard engine.

1. Shim pack
2. Adapter housing
3. Oil filler cap
4. Thrust washer
5. Bushing
6. Idler gear
7. Cap screw
8. Idler shaft
9. Shaft bearing
10. "O" ring
11. Locking bolt
12. Front bushing
13. Drive gear
14. Rear bushing
15. Quill

gear (6) and pump drive gear (13). Keep shim pack (1) together. To disassemble the drive assembly, first remove locking bolt (11) and, working through front of housing, push drive gear (13) and bushing (14) rearward out of housing. The 0.998-0.999 drive gear hub should have a diametral clearance of 0.0008-0.0025 in bushing (12). Renew gear and/or bushing if clearance is excessive, or if gear or bushing are scored or otherwise damaged. To remove the idler gear (6), bushing (5), thrust washers (4) and idler shaft (8), remove cap screw (7) and, using a long punch inserted into the cap screw hole, tap the shaft (8) and shaft bearing (9) rearward out of bushing (5) and housing (2). Withdraw gear (6), thrust washers (4) and bushing (5). The 0.748-0.749 O.D. idler bushing (5) should have a 0.0008-0.0022 diametral clearance in the gear hub and 0.0002-0.0017 diametral clearance on shaft (8). New thrust washers (4) have a thickness of 0.066-0.068.

To reassemble, remove the nut and flat washer from the rear end of shaft (8) and withdraw shaft bearing (9) and "O" ring (10) from shaft. Assemble idler gear (6), bushing (5) and the forward of the two thrust washers (4) in the adapter housing (2), and insert the shaft (8) through idler gear bushing and washer. Start the retaining cap screw (7) in the forward end of idler shaft and tighten sufficiently to retain the shaft and gear. Install rear thrust washer (4) and bearing (9), with "O" ring in place, reinstall shaft nut and tighten nut and retaining cap screw (7).

Backlash between idler gear (6) and the camshaft gear should not exceed 0.006. To measure the backlash, place a small piece of 0.003 thickness shim paper over one of the exposed idler gear teeth. Using new gaskets in shim

pack (1), install the adapter housing on the engine block and tighten the retaining cap screws securely. Rotate the engine just enough to pass the paper through the gear teeth, remove the adapter and examine the shim paper. If the paper shows a heavy impression but is not punctured, the backlash is within limits. If the backlash is excessive, adjust by means of steel or paper shims in shim pack (1). Steel shims are available in thickness of 0.005 and 0.010, and paper gaskets in thicknesses of 0.006 and 0.012. Use one paper gasket of the proper thickness on each side of shim pack.

After installing the housing and idler gear, install and time the injection pump drive gear (13) as follows:

Crank the engine until No. 1 piston is coming up on the compression stroke. Insert a ¼-inch rod in the timing hole on the left side of the flywheel housing, and continue cranking until the rod aligns with, and slides into, a corresponding hole in the front face of the flywheel. This positions the crankshaft at the initial timing position of 16 degrees BTDC for No. 1 piston. Now insert pump drive gear into the adapter housing so that, when completely seated, the master spline is inclined at an angle of 45 degrees toward the engine block as shown in Fig. MF269. When drive gear is properly positioned, install adapter bushing (14—Fig. MF268) so that locating hole aligns with the threaded hole for bolt (11) and install the bolt.

CAMSHAFT

73. To remove the camshaft, it is necessary to remove the timing sprocket cover as outlined in paragraph 70, the rocker arms and shaft assembly and push rods as outlined in paragraph 69 and the timing chain and sprockets as outlined in paragraph 71. Then proceed as follows:

Disconnect high pressure lines to nozzles at injection pump and remove injection pump and drive housing. Disconnect two fuel lines to transfer pump and remove pump. Remove inspection cover from left side of cylinder block and block up tappets using clothes pins or rubber bands as shown in Fig. MF271. Remove camshaft sprocket hub retaining nut (42—Fig. MF270) and remove hub (40). Unbolt and remove camshaft front bearing (36). Withdraw camshaft from front of engine.

Center and rear camshaft journals ride in bushings (37 & 38) which are held in place by doweled capscrews (S—Fig. MF271). Removal of same allows bushing replacement. Center and rear bushing bore diameter is 1.6855-1.6873. Front bearing bore is 1.5620-1.5635. Running clearance of camshaft journals is: Front, 0.0025-0.0045; Center and Rear, 0.0010-0.0033.

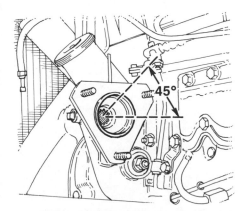

Fig. MF269—Master spline in pump drive gear properly located for installation of injection pump.

Fig. MF270—Standard Diesel camshaft, timing sprockets, timing chain and associated parts.

35. Camshaft
36. Camshaft front bearing
37. Camshaft center bushing
38. Camshaft rear bushing
39. Camshaft sprocket
40. Cam sprocket hub
41. Woodruff key
42. Lock nut
43. Tab washer
44. Crankshaft sprocket
45. Timing chain
46. Chain tensioner assembly
47. Oil pump drive gear
48. Oil deflector

Letter	Piston Dia.
"F"	3.3091-3.3095
"G"	3.3096-3.3100
"H"	3.3101-3.3105

Cylinder sleeves are of the slip-in, dry type and stand out above the block 0.001-0.003. If stand-out is excessive, check for foreign material under sleeve flange. Sleeve bore sizes are as follows:

Letter	Sleeve Bore
"F"	3.3130-3.3135
"G"	3.3135-3.3140
"H"	3.3140-3.3145

Recommended piston skirt clearance of 0.0035-0.0044 is best determined by using micrometers on the mating parts. Maximum allowable cylinder wall taper is 0.008.

There are five piston ring grooves, of which only four are utilized (above the piston pin). The top compression ring is not interchangeable with the second and third compression rings. The fourth groove is occupied by a scraper type oil ring. Compression ring side clearance is 0.0032-0.0052. Piston ring gaps are as follows: Top, 0.009-0.017; Second and Third, 0.009-0.017; Fourth 0.018-0.037. When installing the compression rings, make sure side marked "TOP" faces up.

Camshaft journal diameters are as follows: Front, 1.5590-1.5595; Center and Rear, 1.6840-1.6845. The camshaft has an end play of 0.0020-0.0075 which is not adjustable.

When reinstalling camshaft, align oil holes in front bearing with oil holes in cylinder block. Oil holes are off-center and alignment is possible in cne position only.

Assemble in reverse of removal and set valve tappet gap as outlined in paragraph 64. Install and time the injection pump as in paragraph 138.

ROD AND PISTON UNITS

74. Connecting rod and piston units are removed from above after removing cylinder head and oil pan.

Before removing the units, however, it is advisable to install capscrews and washers as shown in Fig. MF272 to prevent sleeves from moving upward with the pistons. Connecting rod and cap correlation marks face away from the camshaft. Combustion swirl impression on piston head faces toward the camshaft.

Connecting rod assemblies are graded and stamped for weight (T-U-X-P-Q-S) in 1½-ounce graduations from 3 lbs., 1 oz. to 3 lbs., 10 oz. When renewing a connecting rod assembly be sure to install an assembly with the same letter mark as previously removed.

Connecting rod bolt torque is 65-70 ft.-lbs.

PISTONS, SLEEVES AND RINGS

75. Aluminum alloy pistons are supplied in three sizes (in steps of 0.005) and are to be matched with a sleeve having the same marking. Piston sizes are:

PISTON PINS

76. The 1.12485-1.12515 diameter full floating piston pins are retained in piston bosses by snap rings and are furnished in standard size only.

Heat piston in hot oil or water when removing or installing pin. Pin should be fitted to a clearance of 0.003-0.005 in the rod bushing bore and a minus 0.0005 to plus 0.0003 in the piston bore.

Fig. MF271 — When removing Standard Diesel camshaft, rubber band can be used to hold tappets up away from camshaft. Camshaft center and rear bushings are retained by dowled cap screws (S).

Fig. MF272—On Standard Diesel, it is advisable to install a sleeve retaining capscrew before removing the piston and rod assembly.

CONNECTING RODS AND BEARINGS

77. Connecting rod bearings are of the shimless, non-adjustable, slip-in precision type renewable from below after removing oil pan.

When installing a new connecting rod assembly, be sure it has the correct letter (weight) marking as noted in paragraph 74. Correlation mark on rod and cap face away from camshaft.

Connecting rod bearings are available in undersizes of 0.010, 0.020, 0.030

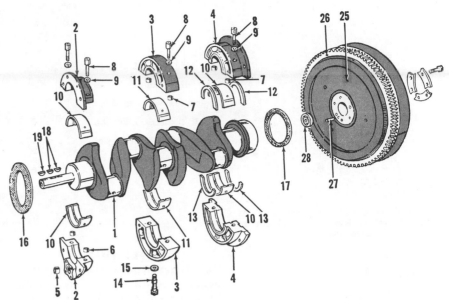

Fig. MF276—Exploded view of Standard Diesel crankshaft, bearing housings and associated parts.

1. Crankshaft
2. Front bearing housing halves
3. Center bearing housing halves
4. Rear bearing housing halves
5. Oil pump housing bushing
6. Dowel bushings, front
7. Dowel bushings, center and rear
8. Allen screws
9. Lock washers
10. Front and rear bearing inserts
11. Center bearing inserts
12. Thrust washers, upper
13. Thrust washers, lower
14. Dowel type cap screw
15. Lock washer
16. Gasket
17. Gasket
18 &19. Woodruff keys
25. Flywheel
26. Starter ring gear
27. Dowel pin
28. Clutch shaft pilot bearing

Fig. MF273 — Standard Diesel crankshaft rear oil seal and seal housing installation.

Fig. MF274 — Standard Diesel front main bearing housing installation.

Fig. MF275 — View of Standard Diesel crankcase showing the doweled capscrew (14) which retains center main bearing housing to block.

and 0.040, as well as standard. Pertinent specifications are as follows:

Crankpin diameter 2.3110-2.3115
Diametral clearance . . . 0.0020-0.0035
Rod side clearance. 0.0065-0.0105
Cap screw torque. . . . . . 65-70 ft.-lbs.

CRANKSHAFT AND BEARINGS

78. The crankshaft is supported in three main bearings which are carried in split cast iron housings. The center and rear bearing housings are installed on crankshaft before fitting same in cylinder block; the front bearing housing, however, is installed after the crankshaft is assembled in the cylinder block.

To renew main bearing inserts, first R&R engine with clutch as outlined in paragraph 62, remove timing cover as in paragraph 70, rocker arm and shaft assembly as in paragraph 69, and timing chain and sprocket as in paragraph 71. Remove cylinder head. Remove clutch assembly as in paragraph 168 and the flywheel assembly as in paragraph 80. Remove oil pan. Refer to Fig. MF273 and remove the crankshaft rear oil seal cover, oil seal and gasket. Remove connecting rod and piston units.

Remove cap screws and Phillips screws retaining front main bearing housing (Fig. MF274) to block and withdraw both of the front bearing housing halves. Remove doweled type cap screw (14—Fig. MF275) which retains the center main bearing housing in the cylinder block; then, withdraw crankshaft, center and rear bearing housing assembly rearward.

Remove Allen screws (8—Fig. MF276) and separate the bearing housing halves. Note the split-type thrust washers (12 & 13) located on each side of the rear bearing housing. These 0.091-0.093 thick thrust washers maintain a crankshaft end play of 0.005-0.012; if end play of 0.012 maximum cannot be maintained, 0.005 oversize thrust washers are available.

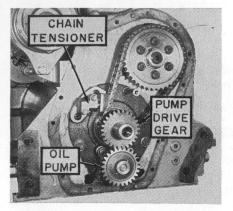

Fig. MF277 — Engine oil pump used on Standard Diesel. Driving gear is keyed to crankshaft.

Fig. MF279 — Left side view of Standard Diesel engine, showing location of oil pressure adjusting screw (AS). Refer to paragraph 82.

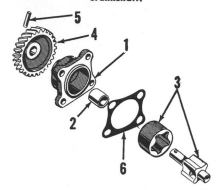

Fig. MF278 — Exploded view of Standard Diesel oil pump and associated parts.

1. Pump body
2. Shaft bushing
3. Rotor set
4. Driven gear
5. Gear pin
6. Gasket

Install copper side of thrust washers facing away from rear bearing housings.

Check the crankshaft journals and main bearings against the values which follow:

Main journal
diameter 2.7515-2.7520
Bearing diametral
clearance 0.0020-0.0035

Installation is the reverse of removal procedure.

Torque the various cap screws to the following values:

Front and rear bearing housings
to block .... 18-20 Ft.-Lbs.
Center bearing housing to
to block .... 39-42 Ft.-Lbs.
Upper to lower bearing
housing .... 25-30 Ft.-Lbs.

CRANKSHAFT REAR OIL SEAL

79. The crankshaft rear oil seal (Fig. MF273) is contained in a cover housing, accessible by separating the engine from the transmission as outlined in paragraph 175. Then remove the clutch assembly and engine flywheel. Remove the cover retaining cap screws and renew the rear oil seal.

FLYWHEEL

80. To remove flywheel, separate the engine from transmission case as outlined in paragraph 175, then remove the clutch unit from flywheel. The starter ring gear can be renewed after removing the flywheel. To install a new ring gear, heat same to 500 degrees F. and install on flywheel with beveled end of teeth facing timing gear end of engine. Flywheel is positioned on crankshaft flange by a locating **dowel**.

OIL PUMP

81. The gyrotor type oil pump has a capacity of 4.74 gallons per minute with 60 psi pressure at 1400 rpm. Oil pump is bolted to the front side of the front main bearing housing and is accessible after removing the timing cover as outlined in paragraph 70.

Remove Woodruff key and slide oil pump driving gear (Fig. MF277) from crankshaft. This will expose the oil pump retaining cap screws and the remainder of removal procedure is evident.

Check the rotors for wear or scoring against the following limits:
Clearance between
rotors 0.0005-0.0040
Rotor wear limit, max. 0.010
Rotor end play, max. 0.004

If rotors are worn, it will be necessary to renew the complete pump. Other components are available separately.

When reassembling oil pump, chamfered edge of outer rotor fits into housing first.

RELIEF VALVE

82. The ball type oil pressure relief valve is located within the oil filter base which is mounted on left front side of cylinder block. The relief valve is set to open at 60 psi. with engine running at 1400 rpm. Opening pressure can be changed by loosening the jam nut and turning adjusting screw (AS—Fig. MF279) as required.

The filter head also contains a spring loaded ball type by-pass valve which permits oil circulation in the event that oil filter becomes clogged.

PERKINS DIESEL ENGINE AND COMPONENTS

R&R ENGINE WITH CLUTCH

83. To remove the engine and clutch as a unit, first drain cooling system and, if engine is to be disassembled, drain oil pan. Shut off the fuel. Remove the hood, disconnect the injector leak off line and both lines from the thermo-start reservoir.

Disconnect radiator hoses and top brace, drag links and radius rods. Support tractor under transmission case,

remove the six cap screws retaining the front support to oil pan and roll the radiator, front support and front axle assembly away from the tractor as a unit.

Disconnect fuel line from shut-off valve and unbolt and remove fuel tank. Remove battery, battery case, and unbolt and remove engine air cleaner and pipes. Disconnect generator wires and wire from thermo-start and disconnect tractormeter cable from drive at rear of camshaft. Disconnect fuel pressure and return lines and throttle control rods from injection pump. Unbolt and remove exhaust pipe. Using the two engine hooks provided, swing the engine in a hoist and unbolt and remove engine from transmission housing.

CYLINDER HEAD

84. To remove the cylinder head, first remove hood, shut off the fuel, disconnect fuel lines and unbolt and remove fuel tank. Drain cooling system, disconnect the upper radiator hose and remove the heat indicator sending unit from water outlet elbow. Disconnect injector lines from injectors and pump adapters, remove leakback lines and unbolt and remove injectors. Disconnect air inlet hose from intake manifold, wire from cold starting unit and external oil feed line from cylinder head and block. Remove intake and exhaust manifolds. Remove

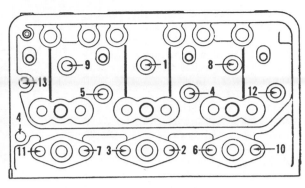

Fig. MF282—Tighten the cylinder head stud nuts to a torque of 55-60 Ft.-Lbs. in the sequence shown. Head studs along left side also serve as injector mounting studs, so a deep socket is required. The rear stud (12) is a waisted stud to prevent installation of similar Perkins P3 head gasket, and must be installed in position shown.

the rocker arm cover and rocker arms assembly, then unbolt and remove the cylinder head.

Mushroom type cam followers operate directly in machined bores in the cylinder head and are prevented from falling out as the cylinder head is raised by the tappet adjusting screw lock nuts.

Head gasket is marked "Top Front" for proper installation. Tighten cylinder head nuts to a torque of 50-60 ft.-lbs., in the sequence shown in Fig.

Fig. MF280—View showing right side of Perkins 3A 152 engine used in MF35 diesel tractor. Separate intake manifold is bolted to right side of cylinder head and is equipped with flame-type cold starting unit. Camshaft is mounted high in engine block and mushroom type cam followers run in bores in cylinder head, eliminating the need for push rods.

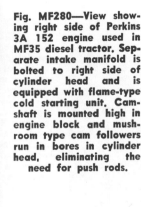

Fig. MF281 — Left side view of 3A 152 engine. Fuel system uses C.A.V. distributor type injection pump and vertically mounted injectors. Spherical combustion chambers are half formed in cylinder head and half by detachable steel caps located on left side of head.

MF282. Head studs along left side of block also serve as injector mounting studs, so a deep socket is required.

VALVES AND SEATS

85. Intake and exhaust valves are not interchangeable and seat directly in the cylinder head. Valve heads and seat locations are numbered consecutively from front of engine as shown at (C) in Fig. MF283. Any replacement valves should be so marked prior to installation. Intake and exhaust valves have a face angle of 44 degrees, a seat angle of 45 degrees and a desired seat width of 1/16-inch. Seats can be narrowed, using 20 and 70 degree stones.

After valves are installed, check the clearance between valve head and gasket surface of cylinder head using the special depth gage (Churchill PD 17A) shown in Fig. MF283, or a straight edge and feeler gage. A minimum clearance of 0.070 and a maximum clearance of 0.140 is allowed. Intake and exhaust valve tappet gap should be set at 0.010 hot. Valve stem diameter is 0.311-0.312 for both intake and exhaust.

VALVE GUIDES

86. The pre-sized intake and exhaust valve guides are interchangeable and can be pressed or driven from the cylinder head if renewal is required. Guides should be pressed into head, using a piloted drift 0.002 smaller than I.D. of guide, until the machined shoulder of guide seats against cylinder head.

Desired valve stem diametral clearance in the 0.314-0.3155 guides is 0.002-0.0045 for both intake and exhaust.

VALVE SPRINGS

87. Springs, retainers and locks are interchangeable for the intake and exhaust valves. Valves are fitted with an inner and outer spring as shown in Fig. MF284. Springs may be installed on the valve with either end up. The inner spring has a shorter assembled

and free length, due to the seating washer (15) and the machined step on spring retainer (12). Renew the springs if they are distorted, discolored, or fail to meet the test specifications which follow:

INNER SPRING
    Approx. free length, inches......$1\frac{3}{8}$
    Lbs. test @ $1\frac{3}{16}$ inches (min.).....8
    Lbs. test @ $2\frac{7}{32}$ inches.........21-25

OUTER SPRING
    Approx. free length, inches......$1\frac{25}{32}$
    Lbs. test @ $1\frac{1}{2}$ inches.......22-23
    Lbs. test @ $1\frac{5}{32}$ inches.......48-52

VALVE TAPPETS

88. The mushroom type tappets (cam followers) operate directly in machined bores in the cylinder head. The 0.6223-0.6238 diameter tappets are furnished in standard size only and should have a diametral clearance of 0.0008-0.0035 in the cylinder head bores.

To remove the tappets after the cylinder head has been removed, first remove the adjusting screw and lock nut, then withdraw the tappet from its bore.

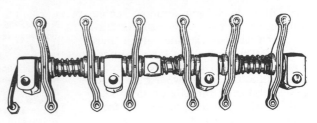

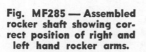

Fig. MF285 — Assembled rocker shaft showing correct position of right and left hand rocker arms.

ROCKER ARMS

89. The rocker arms and shaft assembly can be removed after removing the fuel tank and rocker arm cover. The rocker arms are right and left hand assemblies and should be installed on the shaft as shown in Fig. MF285. Desired diametral clearance between new rocker arms and a new rocker arm shaft is 0.0008-0.0035. Renew shaft and/or rocker arm if clearance is excessive. Rocker shaft diameter is 0.62225-0.62375.

The amount of oil circulating to the rocker arms is regulated by the rotational position of the rocker shaft in the support brackets. This position is indicated by a slot in the front end of

rocker shaft. When the slot is positioned horizontally the maximum oil circulation is obtained. In production, this slot is positioned 30 degrees from the vertical, and the position indicated by a punch mark on the front support bracket. (See Fig. MF285A.) When reassembling, position the rocker shaft as indicated by the punch mark and check the rocker arms assembly for ample but not excessive lubrication.

VALVE TIMING

90. To check the valve timing when the engine is assembled, set the intake valve clearance on No. 1 cylinder to 0.014 (cold), insert a 0.002 feeler gage in the tappet gap, and turn the engine in the normal direction of rotation until a distinct drag is felt on the feeler gage. At this time the crankshaft should be at 13 degrees before top dead center; or 1 43/64 inches before the TDC mark on the flywheel.

TIMING GEAR COVER

91. To remove the timing gear cover, drain cooling system, disconnect headlight wires and remove hood. Disconnect drag links, radius rods and radiator hoses, and unbolt and remove the front support, radiator and axle assembly as a unit.

Fig. MF283 — Measuring valve head clearance with special depth gage (A). A straight-edge and feeler gage may be used. If clearance exceeds 0.140, the valve must be renewed. See text. Cam followers (B) mounted in head can be withdrawn after removing adjusting screws and lock nuts. Valves and seats are numbered in production as shown at (C).

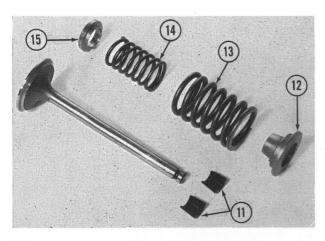

Fig. MF284 — Exploded view of valve, spring, and retainer. Spring seat (15) and retainer (12) have milled steps to apply proper tension to shorter inner spring.

Fig. MF285A — View of rocker shaft and front bracket showing location of punch mark and slot used to mark oil metering adjustment. See text.

Remove water pump and fan assembly and unstake and remove starting jaw and crankshaft pulley. Disconnect and remove the generator and brackets then, unbolt and remove the timing gear cover. The crankshaft front oil seal, located in timing gear cover, can be renewed at this time and should be installed with a suitable driver so that lip faces rear of cover.

TIMING GEARS

92. The timing gear train consists of the crankshaft gear, camshaft gear, pump drive gear and an idler gear connecting the other three gears of the train.

Before attempting to remove any of the timing gears, first remove fuel tank, rocker arm cover and rocker arms assembly to avoid the possibility of damage to the pistons or valve train if either the camshaft or crankshaft should be turned independently of the other.

Timing gear backlash should be 0.003-0.006 between the idler gear and any of the other gears in the timing train. Replacement gears are available in standard size only. If backlash is not within recommended limits, renew the idler gear, idler gear shaft and/or any other gears concerned.

To remove the timing gears or time the engine, unstake and remove the idler gear retaining bolt and slip the gear off the idler shaft. The shaft has a light press fit in the engine block and is further positioned by the locating pin shown in Fig. MF286. Pry the shaft from its place in the block if renewal is indicated.

The crankshaft gear is keyed in place and fits the shaft with 0.001 press fit to 0.001 clearance. If the old gear is a loose fit, it may be possible to pry it off the shaft with a heavy screw driver or pry bar. If a puller is needed, it will first be necessary to remove the oil pan and the small lower section of timing gear housing.

The camshaft gear and injection pump drive gear can be removed by removing the securing cap screws and withdrawing the gears.

To install the gears and time the engine, refer to the appropriate following paragraphs:

93. **CAMSHAFT GEAR.** The gear is attached to the camshaft by cap screws through three equally spaced holes in the gear flange. It is possible to install the gear in three positions, only one of which is correct. One of the attaching holes is marked with a letter "D" stamped on the front face of the gear. The camshaft is marked with a stamped letter "D" on the hub flange near one of the three cap screw holes. The camshaft gear is correctly installed when the two "D" marks are aligned as shown at (X), Fig. MF287.

94. **INJECTION PUMP DRIVE GEAR.** The injection pump drive gear bolts directly to the pump adapter and is carried by the injection pump shaft bearings. To remove the gear after the timing gear cover has been removed, back out the three retaining cap screws and lift off the gear. When reinstalling, align the timing marks as outlined in paragraph 96 and install the pump gear with the idler gear and pump drive gear timing marks aligned.

95. **IDLER GEAR AND HUB.** The idler gear should have 0.001-0.0035 diametral clearance on the 2.123-2.1238 idler hub. The idler gear is retained by the retaining washer and long cap screw which passes through the center of the renewable hub stud and threads into the engine block.

Idler gear is marked with three timing marks which align with the timing marks on camshaft gear, crankshaft gear and injection pump drive gear.

96. **TIMING THE GEARS.** Due to the odd size of the idler gear, the timing marks will align only once in 18 crankshaft revolutions. To time the engine after the gears are removed, rotate the crankshaft until the keyway and timing mark are in a vertical position, then rotate the camshaft and injection pump drive gears until their timing marks are approximately aligned with the center of the idler gear hub. Install the idler gear so that the three timing marks on the idler gear are meshed with the marks on the camshaft, crankshaft and injection pump drive gears. When proper alignment has been obtained, secure the idler gear with the retaining washer and cap screw. Fig. MF287 shows a view of the engine with the timing marks in proper alignment.

TIMING GEAR HOUSING

97. After removal of the timing gears as outlined in paragraph 92, the injection pump as outlined in paragraph 137, and the power steering pump on tractors so equipped, the timing gear housing can be removed as follows:

Fig. MF286—Timing gear idler stud removed, showing locating pin and oil feed holes in block and stud.

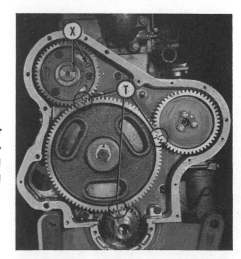

Fig. MF287 — Timing gear train showing timing marks (T) properly aligned. Camshaft gear is correctly installed when stamped "D" marks are in register as shown at (X).

Drain the engine oil and unbolt and remove the oil pan, then unbolt and remove the small section of timing gear housing extending below the crankshaft and serving as a sealing surface for front of oil pan. Unbolt and remove the timing gear housing. The front face of the engine block contains a large expansion plug closing the water jacket and a small expansion plug blocking the front opening of the engine oil gallery.

CAMSHAFT

98. To remove the camshaft, first remove the timing gear cover as outlined in paragraph 91 and the rocker arms assembly as outlined in paragraph 89. Secure the valve tappets in their upper-most position, remove the fuel lift pump, and withdraw the camshaft and gear as a unit.

Camshaft end thrust is controlled by means of a leaf type spring attached to the timing gear cover. Because of the spur type gears used, camshaft end play does not present a problem.

The camshaft journals ride in three machined bores in the engine block. The center journal is pressure lubricated by means of an external oil line (A—Fig. MF288). Oil for lubrication of the rocker arms and valve train is metered by the camshaft journal and fed to the rocker shaft by means of the second external oil line (B). The front and rear camshaft journals are gravity lubricated by the return oil from the rocker arms assembly.

Camshaft journal diametral clearance is 0.004-0.008 for all three journals. Journal diameter is as follows:

Front journal 1.869-1.870
Center journal 1.859-1.860
Rear journal 1.839-1.840

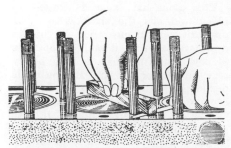

Fig. MF289—Using a straight edge and feeler gage to measure height of semi-finished piston. Piston crown must be milled to provide recommended assembled height of 0.000-0.005 below block face.

ROD AND PISTON UNITS

99. Connecting rod and piston units are removed from above after removing cylinder head, oil pan and rod bearing caps. Cylinder numbers are stamped on the connecting rod and cap. When reinstalling rod and piston units, make certain the correlation numbers are in register and face away from the camshaft side of engine.

Tighten the connecting rod nuts to a torque of 70-80 ft.-lbs., when reinstalling the rod and piston units.

PISTONS, SLEEVES AND RINGS

100. Aluminum alloy pistons are supplied in standard size only, and are available in a kit consisting of piston, pin, rings and sleeve. Note: Piston skirt clearance specifications are not supplied by the manufacturer. As a general rule, ring groove wear will determine the necessity for piston renewal; however, piston renewal due to skirt wear or collapse must be left to the discretion of servicing mechanic.

The chrome plated, thin steel cylinder sleeves should be renewed when visible wear is present at the top of the ring travel, or if the sleeve is damaged in any other way. The 0.0425 thick cylinder sleeves fit 0.001 loose to

0.001 tight in the 3.6875-3.6885 block bores. When installing sleeves, make certain that the block bore, including the counterbore for sleeve flange, and the cylinder sleeve, are absolutely clean and free from any particles of carbon or dirt which might cause distortion of the thin-walled sleeves. Also, examine the sleeves for nicks, burrs or other damage. Apply a light coating of engine oil to the outside of sleeve and press into cylinder bore until the flange is fully seated in the block counterbore. After allowing a few minutes for sleeve to conform to the shape of the cylinder wall, measure for concentricity of sleeve in cylinder bore. Sleeve should measure 3.6015-3.6025 in two directions at top, center and bottom of sleeve. Top flange should be 0.001-0.009 below top surface of the block.

Pistons supplied as parts stock are semi-finished at the piston crown. When renewing the piston, assemble piston, without the rings, on the correct rod and install assembly in the engine block. Turn the piston to top dead center position, lay a straight edge across the top of the piston and measure the height of the piston above the top surface of the block as shown in Fig. MF289. After noting the measurement, remove and disassemble the rod and piston unit, mount the piston in a lathe or milling machine and remove material from the top of the piston equal to the measured amount, plus 0.002. When installed, the crown of the piston should be 0.000-0.005 below the block top surface.

The three 0.093 cast iron compression rings have a side clearance of 0.0019-0.0039 in the piston grooves. Top ring is plain faced and may be installed on the piston either side up. The second and third compression rings are taper faced and marked "T"

Fig. MF288—Right side of engine block showing external oil lines. Oil from main gallery is fed through lower line (A) to center camshaft bearing. Metered oil from camshaft is fed to cylinder head and valve train through upper line (B).

Fig. MF290 — Packing type crankshaft rear oil seal is housed in a two-piece retainer bolted to engine block and main bearing cap as shown.

for correct installation. The two 0.250 oil control rings have a side clearance of 0.002-0.004 in their grooves. All piston rings should have an end gap of 0.009-0.013.

PISTON PINS

101. The 1.24975-1.2500 diameter floating type piston pins are retained in the piston bosses by snap rings and are available in standard size only. The renewable connecting rod bushing must be final sized after installation to provide a diametral clearance of 0.0005-0.001 for the pin. Be sure the pre-drilled oil hole in the bushing is properly aligned with the hole in the top of the connecting rod when installing new bushings. The piston pin should have a thumb press fit in the piston boss bore after piston is heated to 160 degrees F.

CONNECTING RODS AND BEARINGS

102. Connecting rod bearings are of the shimless, non-adjustable, slip-in precision type, renewable from below after removing oil pan and bearing caps. When renewing bearing shells, be sure that the projection engages milled slot in rod and cap, and that the correlation marks are in register and face away from the camshaft side of the engine. Replacement rods should be marked with the cylinder number in which they are installed. Bearings

are available in standard, as well as undersizes of 0.010, 0.020 and 0.030.
Crankpin diameter 2.2485-2.2490
Diametral clearance 0.002-0.0035
Side clearance 0.0095-0.0133
Rod length, C. to C. 8.999-9.001
Rod nut torque . . . . 70-80 ft.-lbs.

CRANKSHAFT AND BEARINGS

103. The crankshaft is supported in four main bearings of the non-adjustable, precision type. Bearing inserts are available in undersizes of 0.010, 0.020 and 0.030, as well as standard. Normal crankshaft end play of 0.002-0.011 is controlled by renewable, flange-type thrust washers located in the rear main bearing cap.

To remove the front main bearing cap, it is first necessary to remove the lower timing gear housing and oil pump. To remove the rear main bearing cap, it is necessary to first remove the engine as outlined in paragraph 83 and remove the clutch, flywheel, and screws retaining rear oil seal retainer to main bearing cap.
Main journal diameter . . . 2.7485-2.7490
Diametral clearance 0.0025-0.0045
Cap screw torque . . . . . . 110-120 ft.-lbs.

ENGINE ADAPTER PLATE

104. The engine flywheel is housed in a cast iron adapter plate which is located on the engine block by two dowels and secured by six cap screws.

To obtain access to the rear main bearing, crankshaft, or rear oil seal, it is first necessary to remove the adapter plate as follows:

Split tractor between engine and transmission as outlined in paragraph 175, and remove clutch and flywheel. Caution: Flywheel is only slightly piloted to crankshaft. Use care when removing securing bolts, to prevent flywheel from falling and causing possible personal injury.

After flywheel is off, remove the cap screws securing adapter plate to engine block and tap plate free of locating dowels.

CRANKSHAFT REAR OIL SEAL

105. The asbestos rope-type rear oil seal is contained in a two-piece seal retainer at the rear of the engine block and rear main bearing cap. The seal retainer can be removed after removing the adapter plate as outlined in paragraph 104. To remove the seal, first remove the two long bolts securing the retainer halves as shown in Fig. MF290, then the cap screws securing retainer to block and bearing cap. Press the rope seal into the retainer halves so that 0.010-0.020 of the seal ends project from the retainer at either end. Fit the retainer halves around the crankshaft and install and tighten the two long bolts through the retainer halves. Using new gaskets, install the retainer and seal assembly on engine block and bearing cap.

Fig. MF292 — Oil pump rotor clearance should not exceed 0.006 when measured with feeler gage as shown.

Fig. MF293 — Rotor to body clearance should not exceed 0.010 when measured with feeler as shown.

RELIEF VALVE

109. The plunger type relief valve is located in the oil pump body and can be adjusted by renewing the spring, to maintain a relief pressure of 50-65 psi at operating speed.

Fig. MF294 — Rotor end clearance should not exceed 0.003 when measured with straight edge and feeler gage.

CARBURETOR

All Gasoline Models

110. Carter and Marvel-Schebler carburetors are used. Carburetor models and their tractor applications are as follows:

	Carter	Marvel-Schebler
TO35 Standard—	UT-2223S	TSX 605
	UT-2418S	
TO35 DeLuxe—	UT-2223S	TSX 605
	UT-2418S	TSX 683
	UT-2612S	
F40—	UT-2418S	TSX 683
MH50—	UT-2418S	TSX 683
	UT-2612S	
MHF202—	UT-2223S	TSX 683
MF50—	UT-2612S	TSX 683
MF202—	UT-2223S	TSX 683
	UT-2612S	
MF35 & MF204—	UT-2223S	TSX 683
	UT-2612S	

Carburetor specifications and overhaul data are as follows:

CARTER
UT-2223S
Float setting 11/64-inch
Initial idle setting...........½-1 turn
Initial load setting......1¾-2¾ turns
Repair kit 1782
Gasket set 259
Inlet needle and seat........25-233S

UT-2418S
Float setting 11/64-inch
Initial idle setting.......½-1½ turns
Initial load setting.......¾-1½ turns
Repair kit 1840
Gasket set 259
Inlet needle and seat........25-233S

UT-2612S
Float setting 11/64-inch
Initial idle setting........¼-1¼ turns
Initial load setting.......¾-1½ turns
Repair kit 1922N
Gasket set 312
Inlet needle and seat........25-248S

MARVEL-SCHEBLER
TSX 605
Float setting ....¼-inch
Initial idle setting............1 turn
Initial load setting........1-1¼ turns
Repair kit 286-1071
Gasket set 16-634
Inlet needle and seat........233-536

FLYWHEEL

CAUTION: Flywheel is only slightly piloted to crankshaft. Use care when removing securing bolts, to prevent flywheel from falling and causing possible personal injury.

106. To remove the flywheel, first separate the engine from transmission housing and remove clutch. Observing the precautions outlined above, unbolt and remove flywheel.

The starter ring gear can be renewed after the flywheel has been removed. To install, heat the new ring gear to 475-500 degrees F. and fit on flywheel with beveled end of teeth facing timing gear end of engine.

Flywheel is secured to crankshaft flange by six evenly spaced cap screws. To properly time flywheel to engine during installation, be sure that unused hole in flywheel aligns with untapped hole in crankshaft flange. Tighten the retaining cap screws to a torque of 75 ft.-lbs. The maximum allowable flywheel runout is 0.008.

OIL PAN

107. The heavy cast-iron oil pan serves as the tractor frame and attaching point for the tractor front support. To remove the oil pan, first support the tractor under the transmission housing and remove the cap screws securing the front support to the oil pan. Remove the cap screws securing the oil pan to transmission housing, then unbolt and remove the pan.

OIL PUMP

108. The rotary type oil pump is mounted on the front main bearing cap and driven from the crankshaft timing gear through an idler gear.

To remove the oil pump, first remove the oil pan as outlined in paragraph 107. Remove the small lower section of the timing gear housing which seals the front of the pan by removing first the two vertical cap screws which fasten the housing section to the block, then the three cap screws at the lower edge of the timing gear cover. Remove the snap ring retaining the idler gear to its shaft and slide the gear forward to obtain access to the pump mounting bolts. Disconnect the pump pressure and inlet lines, then unbolt and remove the pump from the front main bearing cap.

Disassemble the pump, clean the parts in a suitable solvent and examine for wear or broken parts. Reassemble the rotors in the pump body and check the clearance between the rotors as shown in Fig. MF292. Clearance should not exceed 0.006. Check clearance between rotor and pump body as shown in Fig. MF293. Clearance should not exceed 0.010. Check clearance between top of rotors and surface of pump body with a feeler gage and straight edge as shown in Fig. MF294. Clearance should not exceed 0.003. The drive and driven rotors are serviced only as a matched assembly. The pump body is serviced only in the complete pump assembly. Renew the front cover plate if the machined surface is worn or scored. The oil pump idler gear is fitted with a bushing which should be renewed if the clearance is excessive.

TSX 683

Float setting .... ¼-inch
Initial idle setting........... 1 turn
Initial load setting....... 1-1¼ turns
Repair kit 286-1146
Gasket set 10-094
Inlet needle and seat........ 233-536

LP-GAS SYSTEM

The model MF50 tractor is available with a factory installed LP-Gas system, using Zenith equipment. Like other LP-Gas systems, this system is designed to operate with the fuel supply tank not more than 80% filled (16 gallons). It is important when starting the engine to open the vapor or liquid valves on the supply tank SLOWLY; withdrawal valves are equipped with excess flow valves. These valves close when volume of flow increases beyond safe limits, shutting off fuel flow.

CARBURETOR

The Zenith model GO-12157 carburetor has three points of adjustments, as follows:

111. MAIN FUEL ADJUSTMENT. The initial main fuel adjustment is made by turning the main adjusting screw (3—Fig. MF303) finger tight to the fully closed position, then backing it out 7 turns. CAUTION: Because an excessively lean mixture reduces power and overheats the engine, the main adjustment should never be set less than 6 full turns open.

When the engine is warm, make a final adjustment under load as outlined in paragraph 114. Tighten the main adjusting screw lock nut when the selected position is obtained.

112. IDLE MIXTURE ADJUSTMENT. Turn the idling mixture adjustment screw (2—Fig. MF303) out

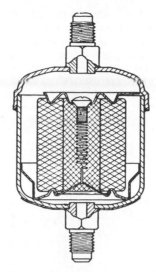

Fig. MF304—Sectional view of MF50 LP-Gas filter.

approximately one turn from the closed position or until the engine idles smoothly when warm.

113. IDLE SPEED ADJUSTMENT. Turn the adjusting screw (1) on the throttle shaft arm in or out to obtain an idle speed of 450 rpm.

114. FINAL ADJUSTMENT. When engine is warm, make the final adjustment to provide good performance under load. Assure that the initial main adjustment has been completed as in paragraph 111.

Final adjustment should be made under load. To make the adjustment, attach a pto dynamometer to the tractor, set governed speed at 1500 rpm and dynamometer load at 20 hp (approximately ⅔ maximum load); then, quickly open throttle to full governed speed. Engine should respond quickly and smoothly. If it does not, open main adjusting screw 1/16 turn at a time and repeat test until engine responds properly.

If a dynamometer is not available, start engine, open throttle and disconnect any three spark plug wires. Turn main adjusting screw until highest engine speed is obtained.

Recheck idle speed and mixture settings and adjust as necessary.

FUEL TANK AND LINES

115. SERVICING. The pressure tank is fitted with fuel filler, vapor return, bleed, pressure relief and liquid and vapor withdrawal valves which can only be serviced as complete assemblies. Before renewal is attempted of any of these units, drive the tractor

to an open area and allow the engine to run until the fuel is exhausted, then open the bleed valve and allow any remaining pressure to escape. The float-type fuel gage assembly consists of a dial face which can be renewed at any time, and a float unit, which can only be renewed if the tank is completely empty. The safety relief valve is set to open at 312 psi to protect the fuel tank against excessive pressures. UL regulations prohibit any welding or repair on LP-Gas tanks. In the event of defect or damage, the tank must be renewed, rather than repaired. Fuel lines and components may be removed at any time without emptying tank if liquid and vapor withdrawal valves are closed and the engine allowed to run until the fuel is exhausted in lines and filter.

FILTER

116. The A-P Controls Division Model 408 filter used in this system is installed in the fuel line between the fuel tank and vaporizer and positioned between the fuel withdrawal valves. It is the "throw-away" type, containing a fibrous filter element as shown in Fig. MF304.

To renew the filter, close both fuel withdrawal valves and allow engine to operate until all fuel in lines, vaporizer, regulator and carburetor is burned out and engine stops. Turn off ignition switch.

Remove three sheet metal screws retaining left side panel to instrument panel and swing out slightly to provide working clearance. Remove two flare fittings at the "tee" in the filter top and a single fitting at the bottom. Refer to Fig. MF305. Remove the two "tee" fittings from the filter and install them on a new filter.

Installation is the reverse of removal procedure.

Fig. MF303—View showing the MF50 LP-Gas carburetor.

1. Idle speed adjustment	2. Idle mixture adjustment	3. Main fuel adjustment

Fig. MF305—View showing installation of MF50 LP-Gas filter. Servicing consists of renewing the complete filter unit.

VAPORIZER

The vaporizer used on the MF50 LP-Gas tractor is a Zenith model No. A965A-21.

117. OPERATION. The vaporizer assembly shown in Fig. MF306 serves the combined purpose of converting liquid fuel from the pressure tank to a vapor, and reducing the higher, varying tank pressure to a constant 10 psi regulated pressure. Fuel, entering as a liquid through inlet passage (37), passes through inlet orifice (38) past inlet valve (13) into the cold chamber (33) where, due to the reduction in pressure, the liquid vaporizes. Inlet valve (13) is initially held open by the pressure of spring (3), acting on the forked piston (10) which is attached to the lower side of diaphragm (9). From the cold chamber (33) the fuel is discharged through swirl passages (35) into the vaporizing chamber (43), where the cooling action associated with the sudden vaporization of the liquid fuel is passed through the heat exchanger shell (25) into the engine cooling liquid in which the unit is mounted. As the vaporizer pressure increases to the pre-determined regulated pressure setting, the vapor pressure, acting through orifice (31)

against the underneath side of diaphragm (9) overcomes the pressure of adjusting spring (3) to close the inlet valve (13) preventing the entry of additional fuel. As the vaporized fuel passes through outlet passage (40), through the secondary regulator and on to the carburetor, vaporizer pres-

sure again drops below the pre-set regulated pressure. The adjusting spring (3) again overcomes the vapor pressure below diaphragm (9), the inlet valve (13) is opened and liquid fuel again enters the vaporizer to maintain the regulated pressure.

118. R&R AND OVERHAUL. Before disconnecting any lines, be sure that all fuel is burned out of the lines, vaporizer, regulator and carburetor by closing the tank withdrawal valves and allowing engine to run until it stops. Turn off ignition switch. Raise hood, drain cooling system completely and remove the right side panel. Disconnect the vaporizer to regulator connection and the tank to vaporizer hose. Remove the vaporizer retaining cap screw (30—Fig. MF306) at underside of water outlet housing and withdraw the vaporizer assembly.

Refer to Fig. MF307 and disassemble as follows: Remove heat exchanger (25) from vaporizer body by removing retaining screw (27). Fibre washer (26) and "O" rings (23 and 24) can be renewed at this time.

Remove inlet orifice retainer (37) and washer (20). Remove the inlet orifice (38) by loosening the lock nut (2) and turning pressure adjustment (1) down as far as it will go by hand; then, threading a ¼—20 inch standard screw into the inlet orifice, grasp the screw body with pliers and tap the inlet orifice out of vaporizer body as shown in Fig. MF308. Note: Do not attempt to turn while withdrawing, since this part is located with a dowel pin, which may shear. Inspect fuel inlet orifice tip (38—Fig. MF307) for nicks or scratches. The tip must be smooth and the seat contact surface parallel with the orifice body. Remove the fibre washer (22) with a scribe or wire hook.

Back out pressure adjustment screw (1) to release spring tension, then remove from diaphragm cover (5). Remove regulator spring (3). Remove the diaphragm cover screws and the diaphragm cover.

Separate the edge of the diaphragm (9) from the vaporizer body (16). Grasping the diaphragm and diaphragm plate (8), twist back and forth gently until piston (10), "O" ring (11), diaphragm and diaphragm plate can be removed from vaporizer body as an assembly. Refer to Fig. MF309. Note: Do not loosen diaphragm assembly screw (6—Fig. MF307) unless diaphragm is defective and needs renewing.

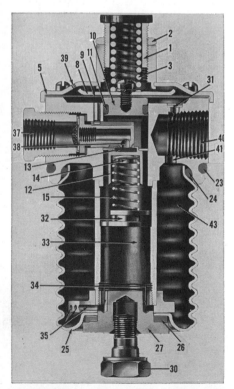

Fig. MF306 — Cross-sectional view of a typical vaporizer assembly used in the MF-50 LP-Gas system. A slight difference in the vaporizer body may be evident upon examination.

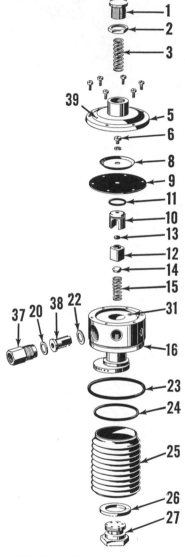

Fig. MF307—Exploded view of the vaporizer assembly. Legend also applies to Fig. MF306.

1. Adjustment screw	25. Heat exchanger shell
2. Lock nut	26. Fibre washer
3. Regulator spring	27. Retainer screw
5. Diaphragm cover	30. Assembly bolt
6. Assembly screw	31. Vent
8. Diaphragm plate	33. Cold chamber
9. Diaphragm	34. Outlet channels
10. Piston	35. Swirl discharge channels
11. "O" ring	37. Fuel inlet
12. Seat retainer	38. Inlet orifice
13. Inlet valve seat	39. Vent
14. Spring button	40. Vaporizer outlet
15. Retainer spring	41. Vapor opening
20. Washer	43. Chamber
22. Fibre washer	
23. "O" ring	
24. "O" ring	

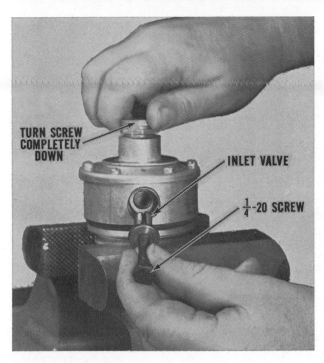

INLET VALVE

¼-20 SCREW

Fig. MF308—Using a ¼ x 20 screw to remove the vaporizer inlet orifice. Pressure adjusting screw must be turned completely down during this operation.

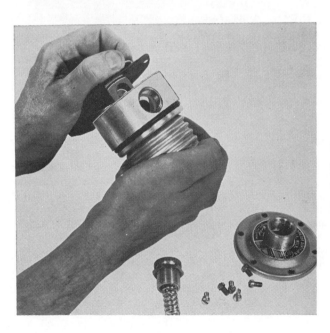

Fig. MF309 — Removing the diaphragm and piston assembly from the vaporizer body. Refer to text.

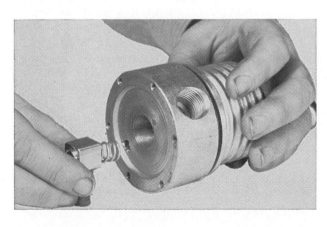

Fig. MF310 — Installing the valve seat retainer, valve seat and spring into the MF50 LP-Gas vaporizer.

119. If diaphragm (9) shows signs of deterioration or cracking, renew as follows: Loosen and remove diaphragm assembly screw (6), diaphragm plate (8) and diaphragm from piston (10). Install new diaphragm, but leave assembly screw loose enough for diaphragm to turn. Push piston (10) into position in the vaporizer body with the opening in the piston skirt parallel with the inlet orifice channel. With a small straight edge inserted in the channel, hold the piston in this position. Rotate the diaphragm (9) until the holes align with those in the vaporizer body (16); then tighten the assembly screw (6).

With the piston (10) removed from the body, valve seat (13), retainer (12) and valve spring (15) may be removed. If retainer (12) sticks in body, use a small wire hook to pull it out. Note: Do not lose spring button (14).

When completely disassembled, clean metal parts thoroughly using an approved solvent. Dry with an air hose, making sure all drilled passages and vents (31 and 39—Fig. MF306) are open and clean. Use new "O" rings, seals and fibre washers.

120. To reassemble the vaporizer, proceed as follows: Assemble the seat retainer (12), spring (15) and spring button (14) at the retainer end. Refer to Fig. MF310. Slide the parts into the vaporizer body until spring (15—Fig. MF307) rests in the spring base spool which is not a removable part.

Align the seat retainer (12) within the vaporizer body so that one flat surface faces the inlet orifice channel. This position allows spring pressure to be distributed equally to all four corners of the seat retainer by the mating surfaces of the piston (10).

Install the assembled piston and diaphragm in the correct relation as noted in paragraph 119. Place the diaphragm cover (5) over the diaphragm and install the six cover screws.

Install and center the regulator spring (3) on the diaphragm assembly screw (6). Install the pressure adjustment (1) and lock nut (2) as an assembly, by compressing the spring (3) slightly and threading the assembly into the cover (5). Screw the adjustment down 6 to 8 turns until the inlet valve seat (13) has moved below the level of the inlet orifice channel.

Install the inlet orifice (38) and a new washer (22), using the ¼—20 inch threaded screw used in disassembly as a handle. The dowel pin in the body and the slot in the inlet orifice must be in alignment.

Install washer (20) and tighten the orifice retainer (37). Install new "O" ring (24) on heat exchanger (25) and place in position. Install retainer screw (27) and fibre washer (26) and tighten securely.

121. VAPORIZER TEST AND ADJUSTMENT. Before installing vaporizer into water outlet housing, proceed as follows: Plug fuel outlet with a suitable pipe fitting to which a shut-off valve is attached. Connect a 30-lb. pressure gage into the test connection in the vaporizer body. Refer to Fig. MF311. Connect the fuel inlet to a compressed air line having approximately 75 lbs. pressure. Back regulator adjustment out as shown. Turn on air pressure and screw in pressure adjustment gradually. The vaporizer should hold each pressure increase without rising or dropping. At several points in the check, release air by opening the shut-off valve in the fuel outlet fitting. The vaporizer should recover the pressure to the original setting as observed on the test gage.

CAUTION: Make certain test air is dry. Do not permit water or vapor to enter vaporizer assembly.

Final adjustment is made by screwing adjustment clockwise until 10 psi is observed on test gage. Tighten lock nut and remove test apparatus.

Install vaporizer into water outlet housing, then install and tighten capscrew (30—Fig. MF306) securely. Make certain "O" ring (23) is in good condition to prevent water leakage.

Connect the vaporizer to regulator connection and the tank to vaporizer hose. The remainder of assembly is evident.

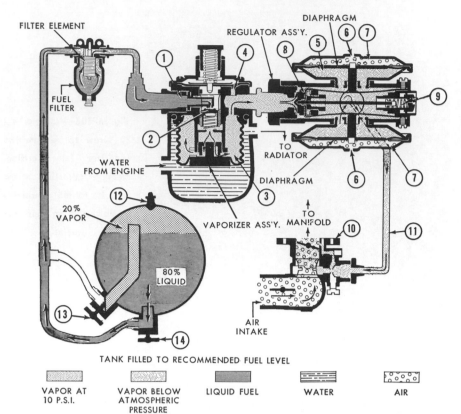

Fig. MF312—Schematic view of a typical LP-Gas system.

1. Liquid fuel inlet	4. Vapor vent	9. Coil spring	12. Relief valve
2. Inlet seat assembly	5. Vapor chamber	10. Carburetor (typical)	13. Vapor withdrawal valve
3. Vaporizing chamber	6. Air vent	11. Vapor pressure line	14. Liquid withdrawal valve
	7. Air pressure		
	8. Valve block assembly		

REGULATOR

The MF50 LP-Gas tractor uses a Zenith model No. B806D-26 regulator.

122. OPERATION. When the engine is turned over, a vacuum condition (slightly less than atmospheric pressure) is set up in the chamber (5—Fig. MF312) of the regulator. This also creates the demand for fuel at the outlet of the vaporizer assembly as discussed in paragraph 117.

Air vents (6) apply atmospheric pressure to the outside of the regulator diaphragm. The difference in the pressure between the inside and outside of the diaphragm causes the two regulator diaphragm assemblies to contract against the leaf spring portion of the valve block assembly (8), lifting the valve block from its seat in the inlet orifice. Vaporized fuel then passes from the vaporizer outlet and into chamber (5) at a pressure slightly less than atmospheric.

When the engine is stopped, the pressure in the chamber (5) rises to atmospheric. The valve block assembly (8) then closes under the force of the coil spring (9).

The system of passages in the carburetor (10) set up a vacuum condition in the hose (11) and chamber (5) that results in the proper flow of fuel to the engine.

The carburetor choke butterfly must be temporarily closed to obtain a richer mixture during starting.

123. R&R AND OVERHAUL. Before disconnecting any lines, be sure all fuel is burned out of the lines,

Fig. MF311 — Using an air hose and pressure gage to test and adjust the MF50 LP-Gas vaporizer. Working pressure is set to 10 psi.

1. Diaphragm vent screen
2. Diaphragm and cover
3. Gasket
4. Regulator body
5. Control valve block and spring
6. Regulator valve seat
7. Fuel inlet orifice
8. Leaf spring lock screw
9. "O" ring
10. Pressure adjusting screw

newed at this time. Ascertain that contact surface of regulator valve seat (6) is smooth and even. Check vent tubes for cleanliness; remove any dirt or obstructions.

To reassemble, insert valve block and leaf spring assembly (5) into regulator body (4). Refer to Fig. MF-314. Note: If resistance is felt, round valve stem may not be centered in the adjusting screw. Refer to Fig. MF315. Remove and repeat until adjusting screw can be turned by hand and slotted head is almost flush with body. Then lightly seat the pressure adjusting screw, turning clockwise. Back out approximately 3 turns.

Insert Zenith Part C161-189 leaf spring gauge set into diaphragm cover bores as shown in Fig. MF316. This will position the leaf springs correctly. Then install and tighten the leaf spring lock screws (8—Fig. MF313).

Install and tighten thoroughly by hand both diaphragm and cover assemblies (2). Install vent tubes toward center of regulator and tighten the attaching fittings. Install the regulator on engine and connect the regulator to carburetor hose and vaporizer to regulator connection. Remainder of assembly is obvious after examination.

vaporizer, regulator and carburetor by closing the tank withdrawal valves and allowing engine to run until it stops. Turn off ignition switch. Raise hood, drain cooling system completely and remove right side panel. Disconnect vaporizer to regulator connection and regulator to carburetor hose. Remove the regulator assembly.

Referring to Fig. MF313, remove the vent tubes and diaphragm and cover assemblies (2), using Zenith Service Tool C161-190 to facilitate removal. Remove fuel inlet orifice (7) and pressure adjusting screw (10). Remove both leaf spring lock screws (8). Remove control valve block and leaf spring assembly (5).

Clean and inspect all parts and renew any which are questionable. "O" ring (9) and gasket (3) can be re-

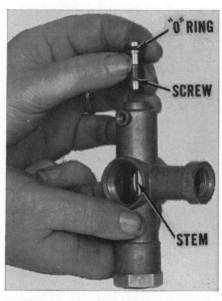

Fig. MF315—Installing the regulator pressure adjusting screw. Stem of valve must enter the hole in the screw.

124. **ADJUSTMENT.** The regulator must be adjusted for slightly below atmospheric pressure regulation as follows: Connect an air hose to the inlet side of the vaporizer unit which has been previously adjusted to 10 psi as outlined in paragraph 121. Ap-

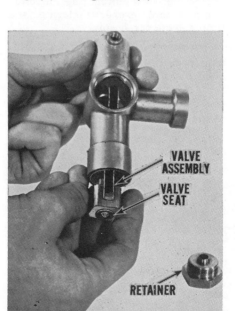

Fig. MF314—Installing valve assembly in the regulator housing. Leaf springs must enter slots at opposite side of bore.

Fig. FM316 — Correctly installing the leaf spring assembly using Zenith Gage C161-189. Refer to text.

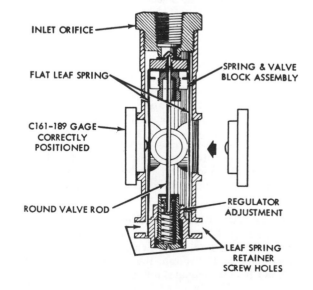

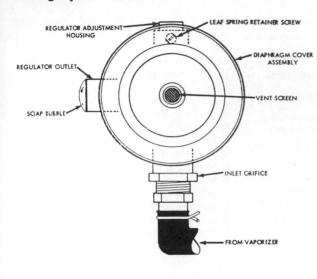

Fig. FM317—Adjustment of the MF50 LP-Gas system regulator.

ply air to the unit, cover the regulator outlet with soapy film solution and slowly turn the pressure adjusting screw (10—Fig. MF313) counter-clockwise until a soap bubble begins to form as shown in Fig. MF317. Then turn the adjusting screw clockwise until the bubble can be maintained without breaking or enlarging. The soap bubble can then be removed and another applied which should lie flat. If bubble lies flat, then turn the adjusting screw ½-turn more clockwise. Add more solution. If the bubble enlarges rather than lies flat, recheck the regulator for improper parts or incorrect assembly; then repeat the procedure.

DIESEL SYSTEM

The TO35 and MF35 diesel fuel system consists of the fuel filters, injection pump and injection nozzles. When servicing any unit associated with the fuel system, the maintenance of absolute cleanliness is of utmost importance. Of equal importance is the avoidance of nicks or burrs on any of the working parts.

Probably the most important precaution that service personnel can impart to owners of diesel powered tractors, is to urge them to use an approved fuel that is absolutely clean and free from foreign material. Extra precaution should be taken to make certain that no water enters the fuel storage tanks. This last precaution is based on the fact that all diesel fuels contain some sulphur. When water is mixed with sulphur, sulphuric acid is formed and the acid will quickly erode the closely fitting parts of the injection pump and nozzles.

125. QUICK CHECKS-UNITS ON TRACTOR. If the diesel engine does not run properly, and the diesel fuel system is suspected as the source of trouble, refer to the Diesel System Trouble Shooting Chart and locate points which require further checking. Many of the chart items are self-explanatory; however, if the difficulty points to the fuel filters, injection nozzles and/or injection pump, refer to the appropriate paragraphs which follow.

FUEL FILTERS AND BLEEDING

126. CIRCUIT DESCRIPTION AND MAINTENANCE. Fuel is drawn from the fuel tank and through the sediment bowl by the primary pump (PP

—Fig. MF318). The sediment bowl assembly (SB), also incorporates a shut off valve. The sediment bowl should be inspected frequently and if water and/or particles of foreign matter are observed, the bowl should be removed and cleaned. The fuel then passes to the first and second stage filters (FS & SS) and on to the injection pump. The primary pump can be disassembled for renewal of gaskets, diaphragm assembly or valve assembly.

The first stage filter contains a replaceable cartridge type element which should be renewed at not more than 500 hour intervals. The filter should be drained at not more than 100 hour intervals. When renewing the element, thoroughly clean the filter interior before installing new element. An air vent at the top of the filter head facilitates air bleeding of the first stage filter.

The second stage filter, next in the fuel flow path, further filters the fuel and also contains a replaceable cartridge type element. This element should be renewed at not more than 1000 hour intervals and the unit drained at not more than 100 hour intervals. Any air in the system at this point is bled back to the fuel tank through a small permanent bleed line in the head. Note: It is important when renewing a second stage filter to also renew the first stage filter, in order to have proper fuel filtering.

The fuel now enters the injection pump where proper metering and timing sequence occurs.

127. BLEEDING. To bleed the system, proceed as follows: Open sediment bowl shut off valve (V). Loosen the air vent on the first stage filter and operate the manual lever on the primary fuel pump up and down fully

Fig. MF318 — Left side view of Standard Engine showing the diesel system components. Refer to text. Perkins engine is fitted with similar components.

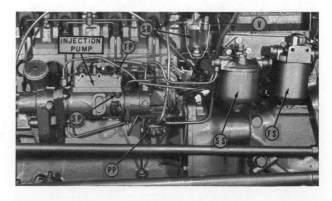

DIESEL SYSTEM TROUBLE-SHOOTING CHART

	Sudden Stopping of Engine	Lack of Power	Engine Hard to Start	Irregular Engine Operation	Engine Knocks	Excessive Fuel Consumption
Lack of fuel	★	★	★	★		
Water or dirt in fuel	★	★	★	★		
Clogged fuel lines	★	★	★	★		
Inferior fuel	★	★	★	★	★	
Faulty primary pump	★	★	★	★		
Faulty injection pump timing		★	★		★	★
Air traps in system	★	★	★	★		
Clogged fuel filters		★	★	★		
Deteriorated fuel lines	★					★
Faulty nozzle		★	★	★	★	★
Faulty injection pump		★	★	★	★	★

until air-free fuel comes out the filter vent. Tighten the vent plug securely. Loosen the vent on the second stage filter; operate the primary pump lever until air-free fuel comes out the vent. Tighten the vent plug.

Loosen the top vent plug (TP) on the injection pump and repeat pumping procedure until air-free fuel flows out the port. Tighten the vent plug. Repeat the procedure with the lower vent plug (SP).

Operate the primary pump lever 10 extra strokes to provide fuel in the bleed back lines to the second stage filter.

Loosen two pressure lines to the injectors and turn the engine over with the starting motor until fuel escapes from the line ends. Tighten the two pressure lines and repeat the procedure with the remaining two pressure lines.

If engine fails to start, it may be necessary to air bleed the system again.

INJECTOR NOZZLES

The 23C Standard diesel engine is equipped with C.A.V. injector assemblies which include Pintaux type nozzle units. Starting with engine No. 86046, a heat shield was added at the combustion chamber end of nozzle. Purpose of shield is to aid in keeping nozzle tip cool to reduce possibility of carbon stoppage of auxiliary hole. Injectors designed for use with heat shields can be installed in earlier engines without shields, but earlier type nozzles cannot be used in later engines.

The Perkins 3A152 engine is equipped with C.A.V. injector assemblies having a closed type nozzle with two 0.013 holes which inject fuel into the combustion area above the piston and into a pre-combustion chamber located in the cylinder head.

WARNING: Fuel leaves the injector nozzles with sufficient pressure to penetrate the skin. When testing, keep your person clear of the nozzle spray.

128. **TESTING AND LOCATING A FAULTY NOZZLE.** If the engine does not run properly and the quick checks outlined in paragraph 125 point to a faulty injector, locate the faulty unit as follows:

If one engine cylinder is misfiring, it is reasonable to suspect a faulty injector. Generally, a faulty injector can

Fig. MF318A.—To completely test and adjust an injector nozzle requires the use of a special tester.

be located by loosening the high pressure line fitting to each injector nozzle in turn, thereby allowing fuel to escape at the union rather than enter the cylinder. As in checking spark plugs in a spark ignition engine, the faulty unit is one which, when its line is loosened, least affects the running of the engine.

Remove the suspected injector unit from the engine as outlined in paragraph 134. If a suitable nozzle tester is available, check the unit as in paragraph 129. If a tester is not available, reconnect the fuel line to the injector assembly and with the nozzle tip directed where it will do no harm, crank the engine with the starting motor and observe the nozzle spray pattern.

If the spray pattern is ragged, unduly wet, streaky and/or not symmetrical or, if nozzle dribbles, the nozzle valve is not seating properly. Send the complete nozzle and holder assembly to an authorized diesel service station for overhaul.

129. **NOZZLE TESTER.** A complete job of testing and adjusting the injector requires the use of special test equipment. Only clean, approved testing oil should be used in the tester tank. The nozzle should be tested for opening pressure, seat leakage, back leakage and spray pattern. When tested, the nozzle should open with a sharp popping or buzzing sound and cut off quickly at end of injection with a minimum of seat leakage and controlled amount of back leakage as outlined in the following paragraphs:

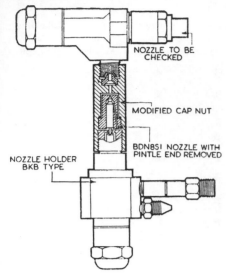

Fig. MF319 — Test apparatus for C.A.V. Pintaux injector nozzle used on TO35 Diesel.

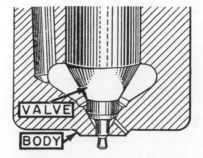

Fig. MF320 — Sectional view of Pintaux type nozzle valve and body.

130. OPENING PRESSURE. The recommended opening pressure is 1900 psi for Pintaux injectors used in the Standard 23C engine, or 1760 psi for closed type nozzles used in Perkins 3A152 engine. To test the pressure, operate tester lever until fuel flows, then install nozzle to be tested. Close the valve to tester gage and apply a few quick strokes to tester lever. If undue pressure is required to operate lever, the nozzle valve is plugged and injector should be disassembled and serviced as outlined in paragraph 135. Open the valve to tester gage and operate tester lever slowly while observing gage reading. If the specified opening pressure is not obtained, remove the injector cap, loosen the adjusting screw lock-nut and turn the adjusting screw either way as required to obtain the recommended pressure. Note: If a new pressure spring is installed, increase opening pressure 80 psi to allow for initial pressure loss.

131. SEAT LEAKAGE. The nozzle valve should not leak at a pressure less than 1500 psi. To check for leakage, actuate the tester lever slowly, and as the gage needle approaches 1500 psi, observe the nozzle tip for drops. Hold the pressure at 1500 psi for ten seconds. If drops appear, or if nozzle tip is wet, the valve is not seating and the injector must be disassembled and overhauled as outlined in paragraph 135.

132. BACK LEAKAGE. If the nozzle seat leakage, as tested above, is satisfactory, check the injector and connections for wetness which would indicate leakage, at the mating surfaces of the nozzle and injector body and at the line connections. If no leakage is found, remove the cap nut and temporarily adjust nozzle opening pressure to 2500 psi; then, operate the tester lever slowly to bring the gage pressure to slightly above 2250 psi, release the lever and observe the time required for the gage pressure to drop from 2250 psi to 1500 psi. For a nozzle in good condition, this time should not be less than 6 seconds. A faster pressure drop would indicate a worn or scored nozzle valve piston and body and the nozzle assembly should be renewed. Note: Leakage of the tester check valve or connections will show up in this test as excessively fast leak-back. If the pressure drop is excessively fast in all injectors tested, the tester, rather than the nozzle, should be suspected.

After completion of the test, reset the opening pressure to the specified setting and reinstall the cap nut.

133. SPRAY PATTERN. The Pintaux nozzle, with which the Standard engine is equipped, is fitted with an auxiliary spray hole as shown in Fig. MF320, to aid starting under cold weather conditions. At starting speeds, pump flow is not sufficient to raise the nozzle valve to clear the pintle hole, and most of the fuel is injected through the auxiliary hole into the hottest part of the combustion chamber. At normal running speeds, however, pump volume is sufficient to raise the nozzle valve clear of the pintle hole, and most of the fuel is injected through the pintle nozzle. Due to the design of the nozzle, special equipment is required to properly check both spray patterns. This equipment consists of a special adapter (C. A. V. No. ET. 846) which is installed on the injector tester between the connector line and the injector to be tested as shown in Fig. MF319. To test the spray pattern, con-nect the special adapter to the tester, open tester gage and set adapter opening pressure to 3225 psi. Close the valve to the tester gage and connect injector to the adapter connection. Operate tester handle at approximately 60 strokes per minute and observe the spray pattern through auxiliary hole. The pattern should be well formed and free from splits or distortions. A slight center core is permitted. Operate tester handle at approximately 140 strokes per minute and observe the main spray pattern. The spray should be well atomized, and free from splits or distortions, a slight center core is permitted.

To test the spray pattern of the two-hole nozzles installed in the Perkins 3A 152 engine, attach the injector to the nozzle tester in the normal man-

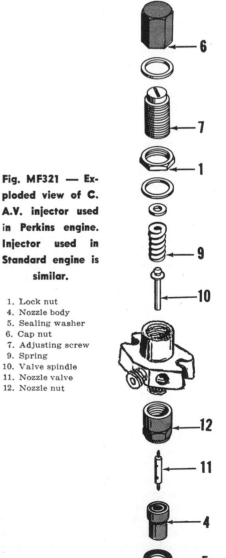

Fig. MF321 — Exploded view of C. A.V. injector used in Perkins engine. Injector used in Standard engine is similar.

1. Lock nut
4. Nozzle body
5. Sealing washer
6. Cap nut
7. Adjusting screw
9. Spring
10. Valve spindle
11. Nozzle valve
12. Nozzle nut

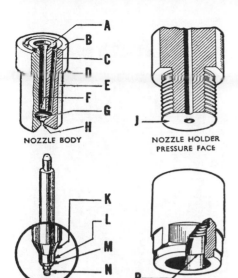

Fig. MF322—Disassembled views of a typical C.A.V. nozzle and holder showing various points for detailed inspection and cleaning.

A. Nozzle body pressure face
B. Nozzle body pressure face
C. Fuel feed hole
D. Shoulder
E. Nozzle trunk
F. Fuel gallery
G. Valve seat
H. Pintle orifice
J. Holder pressure face
K. Valve cone
L. Stem
M. Valve seat
N. Pintle
P. Nozzle retaining shoulder

ner and adjust opening pressure as outlined in paragraph 130. Operate the tester handle at approximately 100 strokes per minute and observe the spray pattern. Two symmetrical sprays should emerge from the nozzle tip. Each should be a finely atomized, misty spray, spreading to about three inches in diameter at about one foot away from the nozzle tip.

If the nozzle fails to pass the spray pattern test, the injector must be disassembled and cleaned as outlined in paragraph 135.

134. REMOVE AND REINSTALL. To remove any one of the injector nozzle assemblies, wash the nozzles, fuel lines and surrounding area with clean diesel fuel to remove any accumulation of dirt or foreign material. On tractors equipped with Perkins 3A 152 engines, remove hood and fuel tank. Unbolt and remove the leak-off line assembly and disconnect the high pressure line from the injector.

After disconnecting the high pressure and leak-off lines, cover open ends of connections with composition caps to prevent the entrance of dirt or other foreign material. Remove the nozzle holder nuts and carefully withdraw the nozzle from cylinder head, being careful not to strike the tip end of the nozzle against any hard surface.

Thoroughly clean the nozzle recess in the cylinder head before reinserting the nozzle and holder assembly. It is important that the seating surfaces of recess be free of even the smallest particle of carbon which could cause the unit to be cocked and result in blowby of hot gases. No hard or sharp tools should be used for cleaning. A piece of wood dowel or brass stock properly shaped is very effective. Do not reuse the copper ring gasket located between nozzle and head; always install a new one. Tighten the nozzle holder screws to a torque of 18-20 Ft.-Lbs.

On the 23C Standard engine after Engine No. 86046, remove the heat shields from the nozzle bore in the cylinder head before cleaning the recess. To install the heat shield, insert the 0.005 shim over small end of shield, pushing shim up to large end. Install the corrugated washer inside the shield with the protrusion up, then slide the assembled shield into nozzle bore of cylinder head. Install injector in the normal manner, using a new copper sealing washer.

Before releasing the tractor for service, start the engine and examine each injector in turn, for blow-by at the copper sealing washer. Correct any observed leakage by loosening one clamp bolt and tightening the other until the condition is corrected.

135. MINOR OVERHAUL (CLEANING) OF NOZZLE VALVE AND BODY. Hard or sharp tools, emery cloth, crocus cloth, grinding compounds or abrasives of any kind should be used NEVER in the cleaning of nozzles. A nozzle cleaning and maintenance kit is available through any C.A.V. Service Agency and other sources.

Wipe all dirt and loose carbon from the nozzle and holder assembly with a clean, lint free cloth. Carefully clamp nozzle holder assembly in a soft jawed vise and remove the adjusting screw protecting cap (6—Fig. MF321). Back off the pressure adjusting screw (7) enough to relieve load from the spring. Remove the nozzle cap nut (12) and nozzle body (4). Normally, the nozzle valve (11) can be easily withdrawn from the nozzle body. If the valve cannot be easily withdrawn, soak the assembly in fuel oil, acetone, carbon tetrachloride or similar carbon solvent to facilitate removal. Be careful not to permit the valve or body to come in contact with any hard surface.

Examine the nozzle body and remove any carbon deposits from exterior surfaces using a brass wire brush. The nozzle body must be in good condition and not blued due to overheating. All polished surfaces should be relatively bright, without scratches or dull patches. Pressure surfaces (A, B & J—Fig. MF322) must be absolutely clean and free from nicks, scratches or foreign material, as these surfaces must register together to form a high pressure joint.

Clean out the small fuel feed channels (C), using a small diameter wire. Insert the special groove scraper into nozzle body until nose of scraper locates in fuel gallery (F); then press nose of scraper hard against side of cavity and rotate scraper to clean all carbon deposits from the gallery. Using seat scraper, clean all carbon from valve seat (M) by rotating and pressing on the scraper.

On Pintaux nozzles, clean the large opening in nozzle tip by using pintle hole cleaner and appropriate size probe, pass the probe down the bore of the nozzle body until probe protrudes through the orifice; then rotate the probe until all carbon is cleared. Clean the auxiliary hole, with a 0.008 probe held in a pin vise, by applying slight downward pressure and rotating probe with the fingers.

Examine the pintle and seat end of the nozzle valve and remove any carbon deposits using a brass wire brush. Use extreme care, however, as any burr or small scratch may cause valve leakage or spray pattern distortion. If valve seat has a dull circumferential ring indicating wear or pitting or if valve is blued, the valve and body should be turned over to an official diesel service station for possible overhaul.

On the closed end nozzles used in Perkins 3A 152 engine, use a 0.013 probe held in pin vise to clean the two injection holes. Use a wooden polishing stick and a small amount of tallow to clean the nozzle valve seat. Clean all carbon and discoloration from nozzle valve with a wire brush. Examine the nozzle and nozzle valve for scoring or wear on piston and seating areas.

Before reassembling, thoroughly rinse all parts in clean diesel fuel and make certain that all carbon is removed from the nozzle holder nut. Install nozzle body and holder nut, making certain that the valve stem is located in the hole of the holder body. Tighten the holder nut. Note: Over-tightening may cause distortion and subsequent seizure of the nozzle valve.

Test the injector as in paragraphs 129 through 133. If the nozzle does not leak under the 1500 psi pressure, and if the spray pattern is satisfactory, the nozzle is ready for use. If the nozzle will not pass the leakage and spray pattern tests, renew the nozzle valve and seat, which are available only in a matched set; or, send the nozzle and holder assembly to an official diesel service station for a complete overhaul which includes reseating the nozzle **valve and seat.**

136. OVERHAUL OF NOZZLE HOLDER. Remove nozzle cap and adjusting screw. Withdraw spring and spindle. Thoroughly wash all parts in clean diesel fuel and examine the end of the spindle which contacts the nozzle valve stem for any irregularities. If the contact surface is pitted or rough, renew the spindle. Renew any other questionable parts.

Reassemble the nozzle holder and leave the adjusting screw lock nut loose until after the nozzle opening pressure has been adjusted as outlined in paragraph 130.

INJECTION PUMP

Both the Standard and Perkins engines are equipped with a C.A.V. distributor type injection pump. A variable speed mechanical governor is built into the pump with regulation up to 2200 rpm.

The subsequent paragraphs will outline ONLY the injection pump service work which can be accomplished without the use of special, costly pump testing equipment. If additional service work is required, the pump should be turned over to an authorized C.A.V. Service Station for overhaul. Inexperienced service personnel should never attempt to overhaul a C.A.V. injection pump.

Standard Engine

137. REMOVE AND REINSTALL. Before attempting to remove the injection pump, thoroughly wash the pump and connections with clean diesel fuel. When any lines are loosened, cover the connection with composition caps to eliminate the entrance of dirt. Close the shut-off valve at sediment bowl. Disconnect all fuel lines from injection pump to injector nozzles. Disconnect two fuel lines from second stage filter. Disconnect fuel shut-off rod and throttle control rod. Remove three nuts retaining injection pump to pump drive housing and withdraw injection pump, taking care not to bend or distort any fuel lines.

The injection pump and drive gear are equipped with a master spline. As long as the pump drive gear is in proper relation to the engine timing gears the pump may be removed and reinstalled without reference to crankshaft timing position.

138. PUMP TIMING. Crank the engine until the No. 1 piston is coming up on compression stroke. Insert a ¼-inch rod in timing hole on left side of flywheel housing and continue cranking slowly until the rod aligns with, and slips into, a corresponding hole in the flywheel. This positions the No. 1 piston in the initial timing position of 16 degrees BTDC.

Shut off the fuel and remove the timing window from the side of the injection pump as shown in Fig. MF323. Loosen the three injection pump mounting bolts and rotate the top of the injection pump until the scribe line on the pump flange is slightly below the scribe line on the adapter housing. At this time the let-

Fig. MF324—Fylwheel timing mark "SPILL 18" centered in timing window for pump timing on Perkins engine.

ter "G" on injection pump rotor should be aligned with the scribed line at the lower hole of the pump snap ring as shown.

When installing the pump, engage the master spline of the pump drive quill with the corresponding master spline in injection pump drive gear, then:

On engines equipped with C.A.V. injection pump Model DPA 3240011 (without automatic advance), rotate top of pump toward engine until scribe lines on pump and adapter flanges are aligned, to obtain the recommended injection point of 17 degrees BTDC.

On engines equipped with C.A.V. injection pump Model DPA 3242645 (with automatic advance), rotate top of pump away from engine until scribe line on injection pump flange is $\frac{1}{16}$-inch below scribe line on adapter flange to obtain the recommended initial injection point of 13 degrees BTDC. The automatic advance mechanism is controlled by the inlet metering valve, to advance the injection timing under partial load operations.

If the proper pump timing cannot be obtained, check the installation of the injection pump drive gear as outlined in paragraph 72 and shown in Fig. MF269.

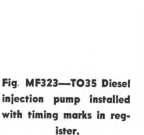

Fig. MF323—TO35 Diesel injection pump installed with timing marks in register.

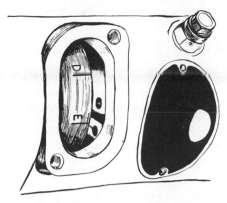

Fig. MF324A—The "E" timing mark on pump rotor should align with scribed line at snap ring lower hole as shown.

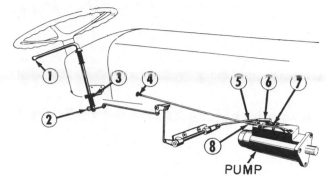

Fig. MF326—TO35 Diesel engine governor linkage.

1. Throttle control lever
2. Clamp bolt
3. Cork washer
4. Fuel shut-off rod
5. Throttle control link rod
6. Throttle control rod clevis
7. Idle adjusting screw
8. High speed adjusting screw

PUMP

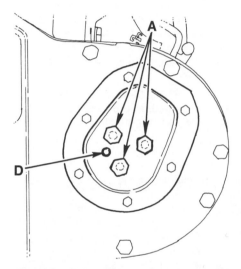

Fig. MF324B — Timing housing cover with oil filler inspection cover removed. Pump timing gear is retained to pump adapter by the three cap screws (A) and timed to pump by the gear dowel (D).

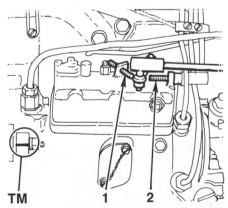

Fig. MF325 — On Perkins engine, pump should be installed with scribe lines (TM) on pump flange and timing gear housing aligned. If pump timing gear is correctly meshed, pump can be removed and reinstalled without regard to crankshaft position. Also shown are idle speed adjusting screw (1) and factory sealed maximum speed adjusting screw (2).

Perkins Engine

139. **REMOVE AND REINSTALL.** The injection pump drive gear is bolted to an adapter secured to the injection pump rotor by an Allen head cap screw, and is carried on the injection pump shaft bearings.

To remove the injection pump, thoroughly clean the pump and connections. Remove the pump to injector pressure lines and disconnect the pump pressure and return lines. Cap all exposed openings to prevent entry of dirt. Unbolt and remove the oil filter inspection cover at the front of the timing gear cover and remove the three cap screws (A — Fig. MF-324B) retaining the drive gear to the injection pump adapter. Be careful not to drop the cap screws or lockwashers into the timing gear case. Disconnect the pump control linkage and unbolt and remove the injection pump.

The injection pump drive gear is timed to the pump shaft adapter by means of a dowel (D) in gear which fits into a slot in the adapter. Although loose in the timing gear housing after removal of the injection pump the drive gear is prevented from getting out of mesh with the idler gear by the walls of the timing gear housing .The injection pump can thus be removed and installed at any time without regard to crankshaft position. Note: Because of possible damage to the loose injection pump drive gear DO NOT turn the engine with the injection pump removed.

When reinstalling pump, rotate the pump adapter so that slot will align with the drive gear dowel and align the scribe mark on the outer mounting bolt flange with the mark scribed on timing gear housing as shown in Fig. MF325, then install and tighten the retaining cap screws. Install and tighten the gear retaining cap screws,

reinstall the inspection cover and complete the installation by connecting the lines and bleeding the system.

139A. **PUMP TIMING.** Crank the engine until the No. 1 piston is coming up on the compression stroke and continue cranking until the "Spill 18" flywheel timing mark is aligned with the center of the timing window as shown in Fig. MF324. Shut off fuel and remove the timing window on the side of the injection pump. The "E" marked scribe line on the pump rotor should now be aligned with the scribed line at the lower hole of the pump snap ring as shown in Fig. MF324A. If it is not, loosen the three pump mounting bolts and rotate the pump in the slotted mounting holes until the marks are properly aligned.

GOVERNOR

All Models

140. The diesel engine governor is integral with the injection pump and only the speed and linkage adjustments will be covered in this section.

With the engine at operating temperature, turn the idle speed adjusting screw (7—Figs. MF326 and MF327) to obtain an engine slow idle speed of 500 rpm.

Loosen the throttle control lever clamp bolt (2) and pull the throttle handle all the way down to the maximum speed position so that the throttle cam is against the bracket assembly stop.

Pull throttle control link rod (5) rearward until the maximum speed stop screw (8) just touches the injection pump stop plate.

While holding the rod in this position, check the engine high idle speed which should be 2200 rpm. If the speed is not as specified, turn screw (8) as required. With rod (5) in the high speed position, securely tighten the rod clamp bolt (2).

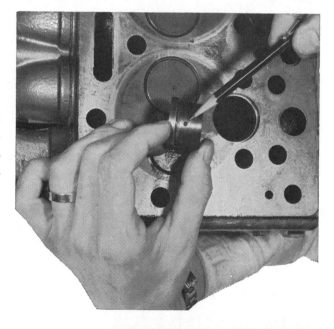

Fig. MF327—TO35 Diesel injection pump installation showing the throttle rod and shut-off rod connections and the speed adjustments.

tered. To remove the pre-combustion chambers, refer to Fig. MF330, loosen the exhaust manifold, remove the three stud nuts retaining the chamber outer half. Tap the cap loose from cylinder head and withdraw cap and copper sealing washer. The outer half of combustion chamber is formed by the cap, while inner half is machined into the cylinder head. Make sure sealing surfaces of head and cap are clean and free from loose carbon or burrs. Renew copper sealing washers when reinstalling caps.

If the throttle control lever tends to "creep", increase spring compression on cork washer (3) by relocating the clamp. If "creeping" tendency remains, renew the cork washer.

PRE-COMBUSTION CHAMBER

141. STANDARD ENGINE. The necessity for cleaning the pre-combustion chambers is usually indicated by excessive exhaust smoking or whenever the fuel economy drops.

To remove the pre-combustion chambers, first remove the cylinder head as in paragraph 63. Remove the injection nozzles, then insert a slightly bent brass rod through nozzle opening and using extreme care, tap the pre-combustion chambers out of head. Pre-combustion chambers are shown installed in Fig. MF328.

The removed chambers can be cleaned with an approved carbon solvent. After cleaning, inspect the chambers and renew any which are cracked or otherwise damaged.

Clean seating surface of chamber and counter bore of cylinder head. When installing, position chamber so that small projection shown in Fig. MF329 will align with groove in head and tap chamber in place using a soft-faced hammer. Chamber must not project beyond gasket surface of cylinder head or head gasket leakage may

occur. If chamber does not seat properly, check for dirt or foreign particles on seating flange.

142. PERKINS ENGINE. Service or cleaning of the pre-combustion chambers is rarely necessary. Inspection and cleaning is advised when excessive smoking exists, fuel economy or power output drops, or when badly carbon fouled injectors are encoun-

COLD WEATHER STARTING AID

Perkins diesel engines, and Standard engines before Tractor Serial No. 188,851, are equipped with the C.A.V. "Thermo-Start" starting aid located in the intake manifold. Standard engines after Tractor Serial No. 188,850 are equipped with individual glow plugs for each cylinder, located in the cylinder head.

143. THERMO-START. The cold weather starting aid, in the intake manifold on right side of tractor,

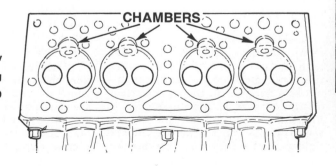

Fig. MF329—Align small projection on pre-combustion chamber with groove in TO35 Diesel cylinder head after cleaning chamber.

CHAMBERS

Fig. MF328—Bottom view of cylinder head showing installation of TO35D pre-combustion chambers

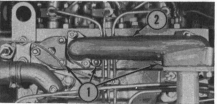

Fig. MF330 — Left side view of cylinder head shownig location of pre-combustion chamber caps on Perkins engine.

1. Caps   2. Exhaust manifolds

serves to pre-heat air prior to its entry into the cylinders.

Turning the starter switch to the "Heat" position, sends electrical current from the battery to the fuel valve solenoid. When the fuel valve opens, fuel from the auxiliary fuel tank flows into coils and becomes vaporized.

Further turning of the starter switch key to the "Heat-Start" position causes engine cranking and the drawing of outside air into the intake manifold. This air mixes with the previously vaporized fuel, becomes ignited and pre-heats the incoming air for easier starting.

144. R&R AND OVERHAUL. The starting aid is a self-contained unit and no service work can be performed. If the starting aid does not function correctly, close shut-off valve, disconnect the fuel line banjo fitting and the electrical lead wire. Remove the unit and install a new one.

Note: The starting aid must be installed with the stamped arrow in the direction of manifold air flow (toward front of tractor) to insure correct operation.

145. GLOW PLUG. The four glow plugs are connected in series with a dash mounted resistor and warning light. From the resistor, the current enters the No. 4 plug, then forward through the other three plugs. The front glow plug is grounded to the engine block.

When the ignition switch is turned to the "Heat" position, the warning light should come on, indicating that current is flowing, heating the four plugs to provide an initial hot spot in each combustion chamber for easier cold weather starting.

The most common form of failure is an open circuit, caused by a burned out plug unit, burned out resistor, or loose connection in the circuit. When an open circuit is present, the warning light will not light and none of the glow plugs operate. To test for the cause of an open circuit, turn the ignition switch to the "Heat" position, and using a screw driver, ground the No. 4 lead in wire connection to the engine block. If the warning light then comes on, the trouble is located in one of the glow plug units. If the light does not come on, the trouble is with the starter switch, resistor unit, warning light, or wiring back to the battery connection. Test each glow plug for a burned out heating element, by using the screw driver to

join the two terminal connections of each plug in turn, while holding the starter switch to the "Heat" position and watching the warning light. If the warning light comes on while shorting out one of the glow plugs, it would indicate that the heating element of that plug is burned out. Another possible cause of failure (or partial failure) would be a short from any of the connections or glow plugs, to the engine block. In the case of a short, all glow plugs on grounded side of short will be inactive, thus causing

heating in none, or only part of the cylinders. To test for a short, connect one lead of a voltmeter to the engine block, and touch the other lead to each plug terminal while holding the starter switch to the "Heat" position. If no reading is shown, that plug is inactive, and the short is to the hot (switch) side of the plug being tested. To remove the glow plugs, disconnect both terminal wires leading to the plug and, using the same care as in removing a spark plug, unscrew and remove the plug.

NON-DIESEL GOVERNOR

MINOR ADJUSTMENT
Non-Diesel Models

146. Warm up engine and adjust carburetor mixture. Disconnect governor to carburetor rod (28—Fig. MF331) at carburetor and adjust engine idle speed stop screw to obtain 400-450 rpm. Reconnect governor to carburetor rod.

Start engine and open hand throttle to wide open position. Desired engine speed of 2175-2225 rpm (2250 rpm for model MF204) can be obtained by loosening the "U" bolt (35) and rotating same on throttle rod (26) until the correct engine speed is reached. Retighten "U" bolt. With engine operating at 1000 rpm check for

surging or unsteady running. If surging exists, check for and remove any binding in the operating linkage. If surging still exists, turn the bumper spring adjusting screw in until surging is eliminated then lock the adjustment.

Note: Engine speeds can be observed on tractormeter.

MAJOR ADJUSTMENT
Non-Diesel Models

147. This adjustment will cover the control linkage if same has been disassembled, or renewed. With the throttle rod (26—Fig. MF331) and the carburetor throttle butterfly valve in the wide open position, adjust rod

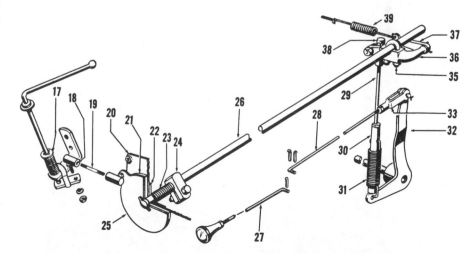

Fig. MF331—Governor controls and linkage for all except TO35 & MF35 Diesel. Engine high idle speed is adjusted by rotating U-bolt (35) with respect to throttle rod (26).

17. Throttle lever assembly	24. Throttle rod clamp	32. Governor lever
18. Ball joint	25. Throttle rod plate assembly	33. Clevis
19. Control link	26. Throttle rod	35. U-bolt
20. Friction plate assembly	27. Choke control rod	36. Linkage rocker
21. Throttle rod plate washer	28. Governor to carburetor rod	37. Compensator spring link
22. Throttle rod washer	29. Lever arm rod	38. Ball joint
23. Throttle rod spring	30. Plunger	39. Compensator spring
	31. Lever spring	

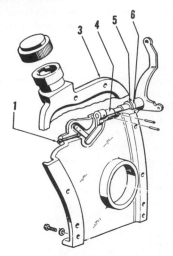

Fig. MF332 — Non-diesel governor control lever and associated parts which are located in timing gear cover.

1. Bumper spring adjusting screw
3. Timing gear cover
4. Needle bearings
5. Oil seal
6. Dust seal

(28) to provide 1/32 inch over-travel. Disconnect lever arm rod (29) at upper end and remove.

Set hand control lever to the idle position and observe if governor compensating spring link is contacting the throttle control rod. If governor compensating spring link (37) is not contacting the throttle rod, loosen the "U" bolt (35) and rotate link (37) until same just touches the throttle rod (26). Retighten "U" bolt.

Reinstall lever arm rod (29) and with the hand control in idle position,

remove governor lever spring (31). Adjust length of rod (29) by loosening the lock nut and rotating the rod in the plunger (30) until bottom of plunger just contacts the governor lever (32) and holds it in the idle position. Reconnect spring and retighten locknut.

Make adjustments for engine speed as listed in MINOR ADJUSTMENT (paragraph 146). If the hand throttle tends to creep, prevent same by loosening throttle rod clamp (24) and compressing spring (23) to increase the friction on the control plate which is located under instrument panel. If this does not remove the throttle creep, renew the cork washer.

FOOT ACCELERATOR
Models MF202-MF204

148. Model MF202 after serial number 301171 is equipped with an accelerator pedal allowing foot operated acceleration to any point between the hand throttle setting and the maximum governed engine speed.

Model MF204 is equipped with a foot operated accelerator pedal, and accelerator linkage coupled to the forward and reverse operating pedals as outlined in paragraph 188.

On each of these models, throttle rod (26—Fig. MF331) is attached at the rear to a throttle arm which is connected by linkage to the accelerator pedal. Throttle rod plate (25) is equipped with a stop pin which limits the movement of the throttle

arm to the idle position, but will not interfere with its movement to the maximum speed position.

To adjust the linkage, proceed as outlined in paragraphs 146 and 147, then adjust the pedal linkage so that maximum governed speed is obtained at the time the pedal contacts the step plate.

R&R AND OVERHAUL
Non-Diesel Models

149. First step in the removal of the governor is to remove timing gear cover as outlined in paragraph 50. The governor operating fork and lever assembly, bearings, and/or oil seal, Fig. MF332, located in the timing gear cover can be renewed at this time.

Remove the camshaft gear retaining nut so as to release the governor driver (balls, cage and cup) assembly.

The governor thrust cup shaft (10—Fig. MF333) has a 0.002-0.004 clearance in the camshaft. It should be renewed if this clearance exceeds 0.006 or if cup is worn at the ball contact surface.

Caution: Check the vent opening in the camshaft shown in Fig. MF255. Vent is located behind the heel of number one cylinder exhaust cam and is drilled to the thrust cup shaft bore.

When reinstalling the upper race assembly, position same so that the lip (stamped projection) is at the 8 o'clock position as shown in Fig. MF333.

Fig. MF333 — Non-diesel governor ball driver and cup installation.

10. Governor race and shaft
11. Nut
12. Ball driver

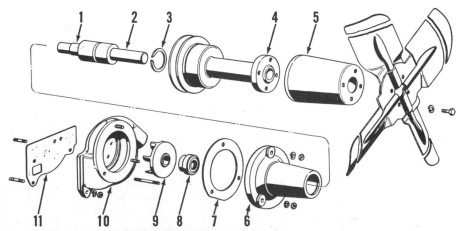

Fig. MF334—Exploded view of non-diesel engine water pump. Pump shaft and bearings are available as an assembled unit only.

1. Slinger
2. Shaft & bearing
3. Snap ring
4. Pulley
5. Shroud
6. Drive shaft support
7. Gasket
8. Seal assembly
9. Impeller
10. Pump body
11. Gasket

Fig. MF335 — Exploded view of Standard Diesel water pump.

24. Bearing housing
26. Bearings
27. Spindle
30. Bearing spacer
31. Snap ring
32. Spindle seal
33. Pulley hub
34. Hub spacer
41. Impeller seal
42. Pump body
43. Gasket
46. Impeller

COOLING SYSTEM

RADIATOR

All Models

150. To remove the radiator, first drain cooling system. Remove hood and grille assembly and disconnect upper and lower radiator hose connections. Remove radiator tie rod on models so equipped and loosen the two cap screws retaining radiator to front axle support. Remove radiator by withdrawing same forward.

THERMOSTAT

All Models

151. Thermostat is located behind the engine water outlet casting and the renewal procedure is evident.

WATER PUMP AND FAN

Non-Diesel Models

152. To remove the water pump, first drain cooling system and remove lower radiator hose. On models F40, MH50 and MF50, remove the radiator. Remove fan blades and fan belt. Remove the three nuts retaining drive shaft support (6—Fig. MF334) to pump body and remove assembly (seal, impeller and shaft). Pump body can be unbolted from engine if desired.

To disassemble pump, remove impeller and pump seal assembly. Remove snap ring (3) and press shaft and bearing assembly (2) toward front and out of shaft support.

Shaft and bearings are available as an assembled unit only.

Standard Diesel Engine

153. To remove the water pump, first drain cooling system and remove the radiator. Remove fan blades, fan belt, fan pulley and hub extension. Remove three capscrews retaining pump to block and one capscrew retaining generator strap to pump bearing housing. Body and block gasket can be renewed at this time.

To disassemble the pump, remove the two remaining bearing housing to pump body capscrews. Separate the sections. Housing to body gasket can be renewed at this time. Remove pulley hub to bearing spindle nut and using a suitable press, press spindle and bearing assembly out of bearing housing. The need for further disassembly is evident.

When reassembling pump assembly to cylinder block, torque to retaining capscrews to 26-28 Ft.-Lbs.

Perkins Diesel Engine

154. To remove the water pump, first drain cooling system and remove hood. Disconnect the upper and lower radiator hoses and radiator brace, then unbolt and remove radiator. Loosen fan belt and remove fan blades. Disconnect the pump outlet and bypass hoses and unbolt and remove the water pump. The top right retaining cap screw is considerably longer than the other three, and all four cap screws are sealed with copper washers. All four cap screws are retained in the pump housing by the fan pulley flange and cannot be removed until the pulley is pressed from pump shaft.

To disassemble the water pump, press the pulley from pump shaft and remove the four mounting cap screws from pump body. Remove the brass backplate and gasket and, using a suitable pilot, press the shaft forward out of the impeller and pump body. The rear seal and slinger can now be removed from the pump body. Pump shaft bearings are sealed, and shaft and bearings are only available as an assembly.

To reassemble the pump, press the shaft and bearing assembly into the pump body until bearing is flush with front of housing. Insert the four mounting bolts in housing, make sure copper sealing washers are all in place and that longer bolt is in upper right mounting hole. Press pulley on shaft until the fan mounting surface of pulley is $5\frac{11}{32}$ inches from gasket surface of pump body. Install the slinger on impeller end of shaft and press a new seal in pump body with thrust face toward the impeller, then press impeller on shaft until rear face is flush with gasket surface of pump body. Reinstall the brass backplate using a new gasket, and complete the assembly by reversing the disassembly procedure.

IGNITION AND ELECTRICAL SYSTEM

DISTRIBUTOR

Non-Diesel Models

155. Delco-Remy distributor models 1112557, 1112570 and 1112585 are used. Specification data follows:

DR 1112557
Breaker contact gap............0.022
Breaker arm spring tension (measured at center of contact), oz...17-21
Cam angle (degrees) 25-34

Advance data is in distributor degrees and distributor rpm.
Start advance ....0-2@275
Intermediate advance ....6-8@750
Maximum advance ......10-12@1050

D-R 1112570
Breaker contact gap............0.022
Breaker arm spring tension (measured at center of contact), oz...17-21
Cam angle (degrees) 25-34

Advance data is in distributor degrees and distributor rpm.
Start advance ....0-2@275
Intermediate advance ......8-10@650
Maximum advance ....11-13@950

D-R 1112585
Breaker contact gap............0.022
Breaker arm spring tension (measured at center of contact), oz...17-21
Cam angle (degrees) 25-34

Advance data is in distributor degrees and distributor rpm.
Start advance ....0-2@225
Intermediate advance ....6.5-8.5@400
Maximum advance ....12-14@550

156. TIMING. To time the distributor, first set the ignition breaker contact gap to 0.022. Crank engine until the number one piston is coming up on compression stroke and continue cranking until the static ignition timing mark on flywheel is in register with the groove in the timing hole as shown in Fig. MF337 or MF338. Loosen the distributor clamp screw and turn distributor until breaker contacts just start to open and tighten the clamp screw.

Running spark timing can be checked with a neon timing light. With engine running at 2200 rpm, the spark should occur 26-30 degrees before top center for engines equipped with a 1112557 distributor. For standard altitude engines equipped with a 1112570 distributor, the running spark should occur 28-32 degrees before top center with engine running at 2200 rpm. For high altitude engines (8.10:1 compression ratio) equipped with a 1112570 distributor, the running spark should occur 22-26 degrees before top center with engine running at 2200 rpm. For LP-Gas engines equipped with 1112585 distributor, the running spark should occur 28-32 degrees before top dead center with engine running at any speed above 1200 rpm. Firing order is 1-3-4-2.

Note: High altitude engines (8.10:1 compression ratio) are identified by a warning plate attached to the engine name plate, two decals attached to the hood and one decal attached to underside of the hood service panel.

GENERATOR, REGULATOR AND STARTING MOTOR

6-Volt System

157. Delco-Remy electrical units are used. Test specifications are as follows:

Generator—D-R 1100530
Brush spring tension (oz.) 16

Field draw
Volts 6.0
Amperes 2.5-2.72

Output (hot)
Maximum amperes 16.0-19.0
Volts 6.9-7.1
RPM 2500

Regulator—D-R 1118308
Cutout relay
Air gap 0.020
Point gap 0.020
Closing voltage (range) 5.9-7.0
Adjust to 6.4

Voltage regulator
Air gap 0.075

Voltage range 6.6-7.2
Adjust to 6.9

Starting Motor—D-R 1109457
Brush spring tension, oz. 24-28

No-load test
Volts 5.7
Amperes 60
RPM 6000

Lock test
Volts 3.3
Amperes 540
Torque (ft.-lbs.) 11.5

12-Volt, Non-Diesel

158. Delco-Remy electrical units are used. Test specifications are as follows:

Generator—D-R 1100998
Brush spring tension (oz.) 16
Field draw
Volts 12.0
Amperes 2.0-2.14

Output (hot)
Maximum amperes 10-12
Volts 14.0
RPM 2800

Generator—D-R 1100362
Brush spring tension (oz.) 28

Field draw
Volts 12.0
Amperes 1.58-1.67

Output (cold)
Maximum amperes 20
Volts 14.0
RPM 2300

Regulator—D-R 1118979
Cutout relay
Air gap 0.020
Point gap 0.020
Closing voltage (range) 11.8-14.0
Adjust to 12.8

Voltage regulator
Air gap 0.075
Voltage range 13.6-14.5
Adjust to 14.0

Regulator—D-R 1118981
Cutout relay
Air gap 0.020
Point gap 0.020
Closing voltage (range) 11.8-14.0
Adjust to 12.8

Voltage regulator
Air gap 0.075
Voltage range 13.6-14.5
Adjust to 14.0

Starting Motor—D-R 1107654
Brush spring tension (oz., min.) 35

No-load test
Volts 10.3
Amperes 75
RPM 6900

Fig. MF337—Static ignition timing marks for all except LP-Gas engines can be viewed through timing hole as shown. For high altitude (8.10:1 CR) engines, the static mark is DC. For standard altitude (6.60:1) engines, the static mark is 6 degrees BTDC.

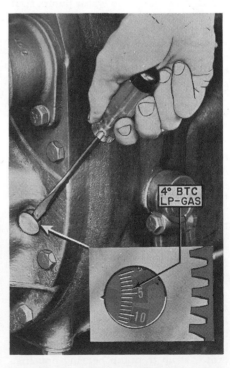

Fig. MF338 — Static ignition timing mark for LP-Gas engine.

Lock test
 Volts 5.8
 Amperes 435
 Torque (ft.-lbs.) 10.5

12-Volt, Diesel

159. Delco-Remy electrical units are used. Test specifications are as follows:

Generator—D-R 1101005
Brush spring tension (oz.) 16
Field draw
 Volts 12.0
 Amperes 2.0-2.14
Output (hot)
 Maximum amperes 10-12
 Volts 14.0
 RPM 2800

Generator—D-R 1100383
Brush spring tension (oz.) 28
Field Draw
 Volts 12.0
 Amperes 1.58-1.67
Output (cold)
 Maximum amperes 20.0
 Volts 14.0
 RPM 2300

Regulator—D-R 1118981
Cutout relay
 Air gap 0.020
 Point gap 0.020
 Closing voltage (range) 11.8-14.0
 Adjust to 12.8
Voltage regulator
 Air gap 0.075
 Voltage range 13.6-14.5
 Adjust to 14.0

Starting Motor—D-R 1108649
         or D-R 1108662
Brush spring tension (oz.) 24
No-load test
 Volts 11.8
 Amperes 40-70
 RPM 6800-9200
Lock test
 Volts 5.85
 Amperes 615
 Torque (ft.-lbs.) 29

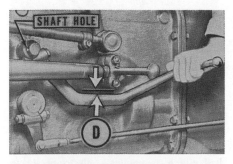

Fig. MF339—Checking the pedal free play adjustment. Free travel (D) is measured between pedal and its stop.

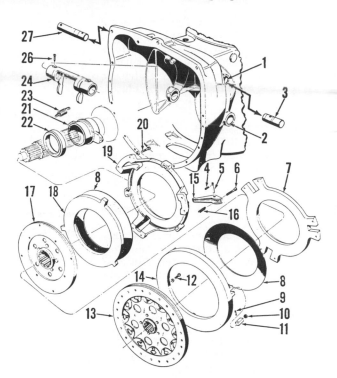

Fig. MF340 — Exploded view of the dual clutch and associated parts used on models TO35 Deluxe, F40 and some MH50 and MF50 models.

1. Clutch release shaft bushing
2. Brake shaft bushing
3. Release pivot shaft
4. Roll pin
5. Release lever pin
6. Adjusting screw
7. Inner pressure plate
8. Belleville clutch spring
9. End rod pin
10. Jam nut
11. Clutch end rod
12. Cap screw
13. Primary lined disc
14. Inner pressure plate
15. Release lever
16. Anchor cup spring
17. Secondary lined disc
18. Outer pressure plate
19. Clutch cover
20. Cover cap screw
21. Release bearing carrier
22. Release bearing
23. Return spring
24. Release fork
26. Set screw
27. Clutch pedal shaft

CLUTCH

This section covers the various clutches used on all models except MF204. Model MF204 is not equipped with a master clutch; instead, a Borg-Warner torque converter is attached to the engine flywheel, and the transmission is equipped with hydraulically activated forward and reverse multiple disc clutches. Service on the torque converter and reversing clutches is covered in the Revers-O-Matic transmission section, paragraphs 187 to 195.

Single Clutch Models

160. Standard model TO35 tractors, some utility model MH50 tractors and some MF50 tractors are factory equipped with a 9 inch Rockford model 9RM, spring loaded, single dry plate clutch which is fitted with a Borg-Warner 165267 cover assembly. Model MHF202 and MF202 tractors are factory equipped with a heavy-duty, 9-inch Rockford spring loaded, single dry plate clutch which is fitted with a Borg-Warner 165385 cover assembly; this same heavy-duty clutch is available for service installation on TO35 Std. tractors, MH50 tractors and MF50 tractors originally equipped with light-duty clutches.

161. **ADJUSTMENT.** Adjustment to compensate for lining wear is accomplished by adjusting the clutch pedal linkage, not by adjusting the position of the release levers on the clutch cover assembly.

To make the adjustment, loosen the bolt clamping the clutch pedal to the clutch release shaft. Insert a rod through the hole provided in the extended end of the clutch release shaft and turn the shaft clockwise until the release bearing just contacts the clutch release fingers. While holding the release shaft in this position, move the clutch pedal with respect to the shaft until there is a distance .(D—Fig. MF339) of ¾-inch between pedal and its stop as shown. Tighten the pedal clamp bolt.

162. **R&R AND OVERHAUL.** The procedure for removing the clutch is conventional after splitting the tractor as outlined in paragraph 174 or 175.

To overhaul the single clutch cover assembly, first mark cover and pressure plate to assure reassembly in same relative position, place cover in a suitable press and remove release lever adjusting screws and pins. Inspect the pressure plate for wear, scoring, or heat cracks on pressure surface and the remainder of clutch parts for wear or damage. Check the spring tension against the values listed below:

Standard duty clutch: Pounds test and test length, 180-190@1$\frac{13}{16}$ inches.

Heavy duty clutch: Pounds test and test length, 140-150@1$\frac{21}{32}$ inches.

To adjust the clutch, reassemble by reversing the disassembly procedure and attach clutch cover to a surface plate or serviceable flywheel, using 0.340 key stock spacers in place of clutch disc. Adjust the fingers to a height of 1 51/64 inches from surface plate for the standard duty clutch, or 1$\frac{35}{64}$ inches for the heavy duty clutch.

Split-Torque Clutch

163. The split-torque clutch is used as a replacement for the dual clutch when excessive clutch work is required. This single plate, spring loaded, dry type clutch is used with a filler plate located in the engine flywheel.

The adjustment and overhaul procedures for this clutch are the same as the heavy duty clutch (165385) covered in paragraphs 160, 161 and 162.

Models TO35 Deluxe (Prior to Ser. 177395; TO35 Deluxe Ser. No. 177520 through 177537)- F40—Some MH50-MF50 (Prior to Ser. No. 515396)

164. The above model tractors are equipped with an Auburn Model 100093-1 dual clutch. Depressing the clutch pedal through its first stage causes the primary pressure plate to retract from its lined disc and interrupt the power flow to the transmission. Depressing the clutch pedal further through its second stage causes the primary pressure plate to retract to a point where it contacts and causes the secondary pressure plate to retract from its lined disc and inter-

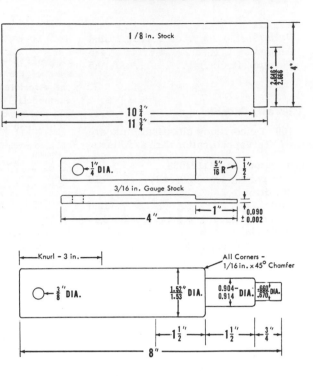

Fig. MF342 — Special tools for installing and adjusting the early type dual clutch assembly. Tools can be made, using the dimensions shown.

rupt the power flow to the hydraulic pump and power take-off input shaft.

165. **ADJUSTMENT.** Adjustment to compensate for lining wear is accomplished by adjusting the clutch pedal linkage, not by adjusting the position of the release levers on the clutch cover assembly.

To make the adjustment, loosen the bolt clamping the clutch pedal to the clutch release shaft. Insert a rod through the hole provided in the extended end of the clutch release shaft and turn the shaft clockwise until the release bearing just contacts the clutch release fingers. While holding the release shaft in this position, move the clutch pedal with respect to the shaft until there is a distance (D—Fig. MF339) of ⅜-inch between pedal and its stop as shown. Tighten the pedal clamp bolt.

166. **R&R AND OVERHAUL.** To remove the clutch, first split tractor as outlined in paragraph 174 or 175 and proceed as follows: Unbolt and remove clutch assembly from flywheel. The primary lined disc (13—Fig. MF340 can be renewed at this time.

Note: In order to maintain the clutch balance when reassembling, carefully punch assembly marks on the clutch cover (19), the outer pressure plate (18) and both of the inner pressure plates (7 & 14). Place the assembly in a press with cover assembly up. Place a bar across the cover and compress the assembly until the re-

lease levers are just free. Loosen the lock nuts and turn the release lever adjusting screws (6) completely out of the clutch end rods (11). Release the compressing pressure and disassemble the remaining parts. Notice that release lever pins (5) are retained by small roll pins (4) driven through the lever pin and into a hole in the clutch cover. To remove the release levers, drive the lever pins (5) out in the direction of the roll pin. This will pull the roll pin from

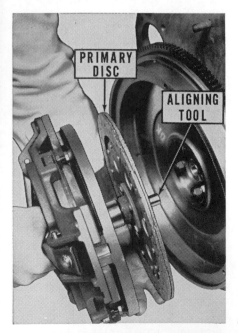

Fig. MF341—Using a special aligning tool to install the primary lined disc and clutch cover assembly on early type dual clutch assembly.

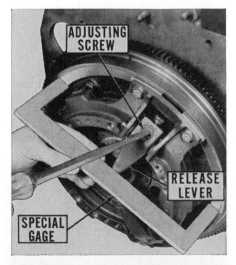

Fig. MF343 — Using special gage to adjust the release lever height on early type dual clutch assemblies. Adjustment should only be made with a NEW primary lined disc installed.

the hole in the clutch cover. Always use new roll pins when reassembling.

When reassembling, proceed as follows: Place the inner pressure plate on a bench with friction face down. Examine the two Belleville-type spring washers (8). If one washer is thicker than the other, install the thicker washer on pressure plate with convex side up. If springs are the same thickness, use either spring. Place pressure plate (7) with flat side up in position on the assembly and align the previously affixed punch marks. Center the secondary lined disc (17) on the pressure plate with the longer hub of the disc up. Position the outer pressure plate on the assembly with flat side toward lined disc and make certain that the assembly punch marks are in register. Install the remaining Belleville spring washer with concave side up, place the clutch cover (19) in position and be sure the assembly punch marks are in register. Using a press and a bar across the clutch cover plate, compress the assembly just enough to start the adjusting screws in the clutch end rods. Tighten the adjusting screws part way and remove the assembly from the press.

With the cover assembly and the primary lined disc (long hub toward clutch cover) on the clutch aligning tool (Fig. MF341), assemble the units to flywheel as shown. With the release lever height gage (Fig. MF342) in the position shown in Fig. MF343, adjust the levers to the gage by turning the adjusting screws in or out as required. Tighten the adjusting screw locknuts when adjustment is complete. Note: This adjustment should be made only with a new primary lined disc in place.

To make certain that the secondary pressure plate begins to release at the proper time, there must be 0.090 clearance between the adjusting cap screws and the secondary pressure plate. With the release levers properly adjusted as previously outlined, loosen the lock nut and turn each of the adjusting screws in or out as required until the 0.090 feeler gage (Fig. MF-342) can just be inserted between the cap screw head and the secondary pressure plate as shown in Fig. MF344. Tighten the lock nuts when adjustment is complete. Note: This adjustment can be made with either a new or used, but still serviceable, secondary lined disc in position.

Model TO35 Deluxe (Serial No. 177395 through 177519; Ser. No. 177538 and Up)-Model MF50 (After Ser. No. 515395)-Model MF35

The above model tractors are equipped with an Auburn combination coil and Belleville spring dual clutch assembly. Depressing the clutch pedal through its first stage retracts the primary pressure plate from its disc and interrupts the power flow to the transmission. Further depressing the clutch pedal through its second stage interrupts the power flow to the hydraulic pump and power take-off input shaft by causing the primary pressure plate to retract against the secondary pressure plate and move it away from its lined disc.

167. CLUTCH LINKAGE ADJUSTMENT. Adjustment to compensate for lining wear is accomplished by adjusting the clutch pedal linkage, NOT by adjusting the position of the release levers on the clutch cover assembly.

To make the adjustment, loosen the bolt clamping the clutch pedal to the clutch release shaft. Insert a rod through the hole provided in the exposed end of the clutch release shaft and turn the shaft clockwise until the release bearing just contacts the clutch release fingers. While holding the release shaft in this position, move the clutch pedal with respect to the shaft until there is a distance (D—Fig. MF-339) of ¾-inch between pedal and its stop as shown. Tighten the pedal clamp bolt.

168. REMOVE AND REINSTALL. To remove the clutch, first split tractor as outlined in paragraph 174 or 175 and proceed as follows: Make-up three special "T" bolts by welding a cross bar to ¼ x 6 inch screws, then add lock nuts. Install the special "T" bolts through the three holes in the clutch cover to hold the assembly together while removing it from the flywheel and tighten the "T" bolt lock nuts to compress the pressure springs. Note: In order to maintain the clutch balance when reassembling, punch assembly marks on the clutch cover, the 9 inch pressure plate, the flywheel plate and the 11 inch pressure plate. Remove the six cap screws securing the cover assembly to the flywheel and remove the clutch as shown in Fig. MF345, using care not to damage or distort the air ring between the clutch assembly and the ventilated flywheel.

169. To install the dual clutch assembly, install the air ring on the flywheel and use two $\frac{5}{16}$ x 3 inch guide studs as shown. Then proceed as follows:

Insert the pilot tool (Fig. MF346) through the driven discs and install the clutch cover into the flywheel.

Note: Be sure that the pressure plate Bellville spring (36—Fig. MF347) is centered in the cover plate (33) when loosening the special "T" bolt lock nuts.

Remove the pilot tool and the three special "T" bolts. If a new clutch cover assembly is to be installed, it is also necessary to remove three cap screws which are used to secure the assembly together when received from stock.

170. OVERHAUL. Note: Dimensions of special tools necessary for servicing the clutch are shown in Fig. MF346. To disassemble a removed dual clutch, proceed as follows: Unhook the three torsion springs (50—Fig. MF347) from the clutch release levers (39). Loosen the adjustment cap screw lock nuts (52) and tighten the cap screws (51) until they bottom against the eleven inch pressure plate (35). Back off the special "T" bolts until it is possible to drive the roll pin (42) down sufficiently to remove the upper release lever pivot pins (41).

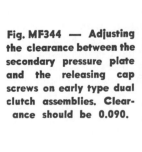

Fig. MF344 — Adjusting the clearance between the secondary pressure plate and the releasing cap screws on early type dual clutch assemblies. Clearance should be 0.090.

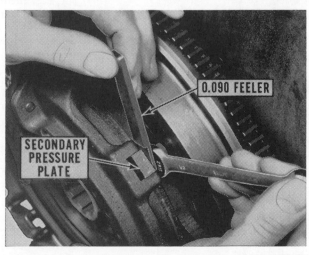

0.090 FEELER

SECONDARY PRESSURE PLATE

Note: Do not attempt to drive roll pins (42) completely out of the cover assembly. To do so will damage the cover assembly or Bellville spring. The pins can be removed very easily once the cover has been removed.

Remove the three upper release lever pivot pins and the torsion springs (Fig. MF348); loosen the lock nuts on the special "T" bolts evenly until the coil springs (55) are fully extended and the clutch cover is free. Then remove the special "T" bolts.

The remainder of the disassembly is evident. The roll pins (42—Fig. MF-347) can be removed at this time.

Coil spring specifications are as follows:

Color ....Yellow
Free length ....2 45/64 inches
Length @ test
load.......1½ inch @ 80-88 lbs.

Thoroughly clean and examine all other parts and renew any which are damaged or worn. Replacement linings are available for both driven plates.

When reassembling, proceed as follows: Place the clutch cover upside down on a bench; then center the Bellville spring (36) in the cover groove with the convex side up. Place the pressure plate (34) on the spring, aligning the previously affixed punch marks. Install the driven disc (37) with the hub side down; place the smooth side of the flywheel plate (38) down on the disc and align the punch marks. Bolt the assembly together temporarily with three $\frac{5}{16}$ x 1½ inch

Fig. MF346 — Special tools for installing and adjusting the late type dual clutch. Tools can be made, using the dimensions shown.

bolts through the cover to flywheel plate mounting holes.

Partially insert the lower release lever pin (54) and coat it with "Lubriplate."

Place the eleven inch pressure plate (35) with friction face down on a bench and install the insulating washers (56) and pressure springs (55). Install the previously assembled clutch cover over the springs, aligning the punch marks. Install the three special "T" bolts until they bottom in the pressure plate (35). Tighten the "T" bolt lock nuts evenly until lower

release lever links (49) can clear the flywheel plate sufficiently to permit inserting of the lower release lever pins (54). Then tighten the "T" bolt lock nuts further to allow insertion of the upper release lever pins (41) through the clutch release levers (39) and torsion springs (50). Secure the upper release lever pivot pins with the roll pins (42) and hook the torsion spring ends. Finally, remove the three temporary $\frac{5}{16}$ x 1½ inch bolts fastening the clutch cover and flywheel plate together. Adjust the unit as follows:

171. INITIAL ADJUSTMENTS. After the clutch assembly has been installed on the engine flywheel, two adjustments are necessary for proper clutch operation:

172. CLUTCH RELEASE LEVER HEIGHT. This adjustment is necessary to insure that the clutch pedal travel will fully disengage the primary and secondary clutch discs. Proceed as follows: When the clutch assembly has been installed, with a NEW primary disc, the release lever height is checked with the special gauge shown in Fig. MF346. The release lever height is measured from the machined surface of the clutch cover to the bearing contacting surface on the adjusting screws as shown in Fig. MF349. Loosen the lock nuts and turn the adjusting screw in or out to obtain the proper height. Tighten the lock nut and recheck the adjustment.

Note: Clutch release lever height must be adjusted with a NEW primary disc installed. If it is desired to use a

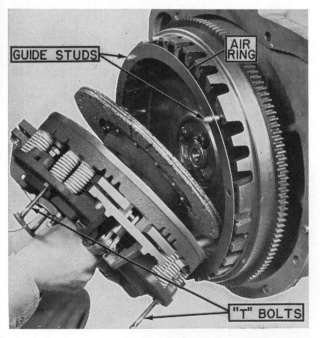

Fig. MF345 — Removal and/or installation of late type dual clutch assembly, showing "T" bolts in place.

1. Transmission case
2. Clutch shaft bushing
3. Brake shaft bushing
4. Release shaft arm
5. Bottom cover
6. Gasket
7. Drain plug
21. Release shaft
22. Release pivot shaft
23. Release bearing fork
24. Release bearing carrier
25. Release bearing
26. Carrier spring
27. Fork screw
33. Clutch cover
34. Pressure plate (9 inch)
35. Pressure plate (11 inch)
36. Belleville spring
37. Driven disc assy.
39. Release lever
41. Upper lever pin
42. Roll pin
46. Adjusting bolt
47. Lock nut
48. Link pin
49. Lever link
50. Torsion spring
51. Bracket bolt
52. Lock nut
54. Lower link pin
55. Pressure spring
56. Washer
57. Drive disc assy.
60. Air ring

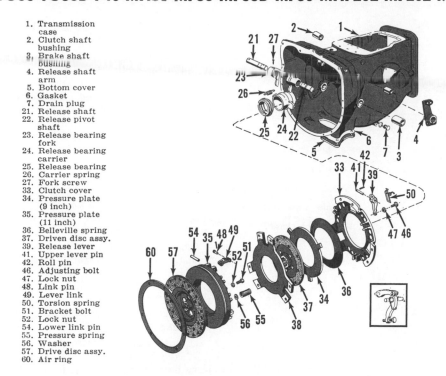

Fig. MF347—Exploded view of the transmission case and late type dual clutch assembly.

partially worn disc, the adjustment should first be made with a NEW disc, then the partially worn disc can be subsequently installed without changing the adjustment.

173. PRIMARY PRESSURE PLATE FREEPLAY. This adjustment determines the point in the clutch pedal travel where the secondary plate begins to release. Proceed as follows: With the release levers properly adjusted as previously outlined, loosen the lock nuts (52 — Fig. MF347) and turn each of the adjusting

screws (51) in or out until the 0.090-inch feeler gage (Fig. MF346) can just be inserted between the cap screw head and the secondary pressure plate as shown in Fig. MF350. Tighten the lock nuts when the adjustment is complete. Note: This adjustment can be made with either a new or used, but still serviceable, secondary lined disc in position.

SPLITTING TRACTOR

All Except TO35 & MF35 Diesel

174. To detach engine from transmission proceed as follows: Drain cooling system and on MF35, TO35, MHF202 and early MF202, tip hood assembly forward and disconnect front end of radius rods and/or drag links. On models F40, MH50 and MF50, remove the cap screws retaining rear end of hood side panels to instrument panel and disconnect drag link from pitman arm. On late MF202 remove hood, and disconnect drag link and radius rods. On all models, shut off fuel and remove fuel line. Unbolt fuel tank from its rear support, loosen fuel tank front support bolts and block up between the fuel tank and the rocker arm cover. Disconnect the heat indicator sending unit from water outlet elbow and cable from starting motor. Disconnect wires from coil and generator, tractormeter cable from gen-

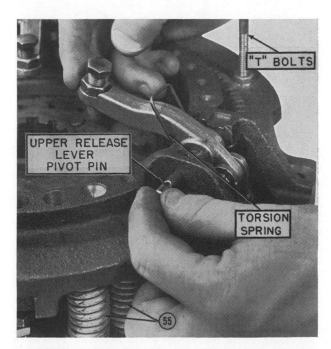

Fig. MF348 — Removing the upper release lever pivot pin.

Fig. MF349 — Adjusting release lever height on late type

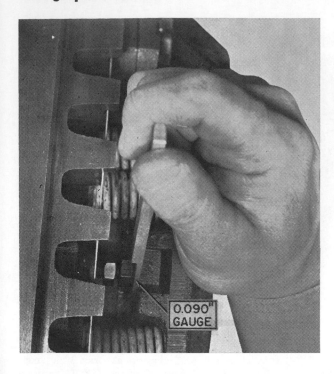

Fig. MF350 — Using the 0.090 inch gage to set the primary pressure plate free-play on late type dual clutch.

Models TO35 & MF35 Diesel

175. To detach the engine from transmission, drain cooling system and tip hood assembly forward. Close the fuel shut-off valve. Disconnect drag links and radius rods at forward end. Remove air cleaner to manifold hose and disconnect starting aid wire. Disconnect both fuel lines from starting aid reservoir.

Disconnect battery cable and wiring harness from generator and solenoid; disconnect heat indicator sending unit from water outlet and unclip from left side of cylinder head. Unbolt exhaust pipe from manifold.

Disconnect fuel line from tank to transfer pump and fuel lines from transfer pump to second stage filter. Disconnect the fuel shut-off control rod, throttle control rod and tractor-meter cable from injection pump. Disconnect oil pressure line banjo fitting.

Unbolt fuel tank from its rear support, loosen the fuel tank front support bolts and block-up between the fuel tank and rocker arm cover. Remove starting motor.

Support both halves of tractor separately. Unbolt the engine from transmission and separate the tractor halves.

erator, choke rod from carburetor and oil gage line from right side of cylinder block. Loosen the U-bolt assembly from front end of throttle rod and disconnect exhaust pipe from manifold. Disconnect battery cables and re-

move battery. Unbolt battery platform from engine. Support engine in a hoist and place a rolling floor jack under transmission case. Unbolt engine from transmission and separate the tractor halves.

TRANSMISSION AND CONNECTIONS

This section covers the transmission used in all models except MF204. Model MF204 is equipped with a Revers-O-Matic transmission consisting of a flywheel mounted torque converter, hydraulically activated forward and reverse multiple disc clutches, and a four-speed sliding gear unit. Service on the Revers-O-Matic transmission is covered in a separate section, paragraphs 187 to 195.

REMOVE AND REINSTALL

All Models Except MF-204

176. To remove the complete transmission housing assembly, first detach (split) engine from transmission as outlined in paragraph 174 or 175 and proceed as follows: Disconnect tail light wires and wires from starter safety switch. Unbolt and remove steering gear housing and transmission top cover assembly. Disconnect brake rods and on models so equipped, remove drag links and radius rods.

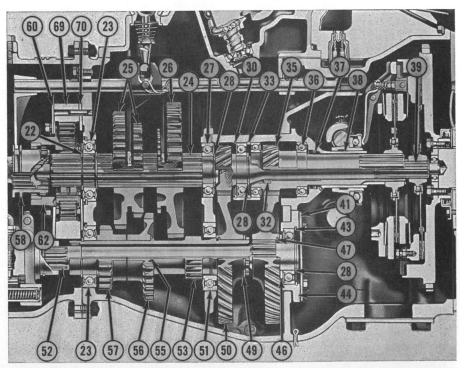

Fig. MF351—Side sectional view of a typical dual clutch and transmission assembly. Transmission shaft bearings are non-adjustable. Notice that the countershaft is hollow to accommodate the live pto front main drive shaft. Refer to legend for Fig. MF357.

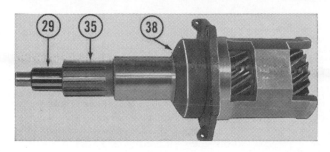

Fig. MF352—Transmission main drive gear (clutch shaft) and retainer assembly used on models with a dual plate clutch.

Support transmission housing and rear axle center housing separately, unbolt and separate the units.

R&R TOP COVER

177. The transmission top cover is integral with the steering gear housing. To remove the unit, proceed as outlined in paragraph 14 for models TO35, MF35, MHF202 and MF202 and paragraph 20 for models F40, MH50 and MF50.

OVERHAUL

178. Data on overhauling the various transmission components are as follows:

179. **MAIN DRIVE SHAFT AND GEAR (CLUTCH SHAFT).** To remove the main drive shaft and gear, first detach (split) engine from transmission housing as outlined in paragraph 174 or 175. Remove set screws retaining the clutch release fork to the shafts, pull the release fork shafts out-

ward and remove the release fork. Then, when working on models with double plate clutch, proceed as outlined in paragraph 180; when working on models with single plate clutch, proceed as outlined in paragraph 181.

180. MODELS WITH DOUBLE PLATE CLUTCH. After performing the work outlined in paragraph 179, proceed to remove the main drive shaft and gear as follows: Remove the front pto housing bearing cap (41—Fig. MF-351) and extract snap ring (28) from forward end of the live power take-off shaft. Use two puller cap screws in the tapped holes provided in the pto shaft front bearing housing (44) and remove the housing and bearing assembly. Remove thrust washer (46). Remove the internal snap ring (47) from the pto main drive gear and withdraw the front pto main drive shaft from transmission case. This allows the drive gear to drop down and clear the transmission main drive shaft retainer.

Unbolt the main drive shaft retainer from transmission case and withdraw the drive shaft and retainer as an assembly (Fig. MF352).

To disassemble the unit, remove snap ring (30—Fig. MF351) and withdraw the main drive gear and shaft. Remove snap ring (33) and bump the pto main drive shaft from the housing. Oil seals (37) in housing and (32) in the pto main drive shaft can be renewed at this time. The need and procedure for further disassembly is evident.

When reassembling, reverse the disassembly procedure and use a protector sleeve or shim stock to avoid damaging the seals when shafts are installed. Shoulder of thrust washer (46) goes toward front.

181. MODELS WITH SINGLE PLATE CLUTCH. After performing the work outlined in paragraph 179, proceed to remove the main drive shaft and gear as follows:

Unbolt the main drive shaft retainer from transmission case and withdraw the drive shaft and retainer as an assembly.

To disassemble the unit, remove snap ring (30—Fig. MF353) and withdraw the main drive gear and shaft. Seal (37) can be renewed at this time. When reassembling, use a protector sleeve or shim stock to avoid damaging the seal when drive shaft is installed.

182. **SHIFTER RAILS AND FORKS.** To remove the transmission shifter rails and forks, first remove the transmission top cover and steering gear housing assembly and proceed as follows:

Disconnect the brake rods, step plates and exhaust pipe. Support both halves of tractor, unbolt transmission from rear axle center housing and separate the units.

Unwire and remove the set screws retaining the selector and shifter forks to the rails. Remove the detent springs and plungers. Unbolt and remove stop plate (16—Fig. MF354) and withdraw the shifter rails and forks from transmission case.

When reassembling, reverse the disassembly procedure. Forks (21 and 21A —Fig. MF355) are interchangeable, but rails for same are not. Rails should be installed with the milled flat toward rear and on top and with the selector lock grooves to the center.

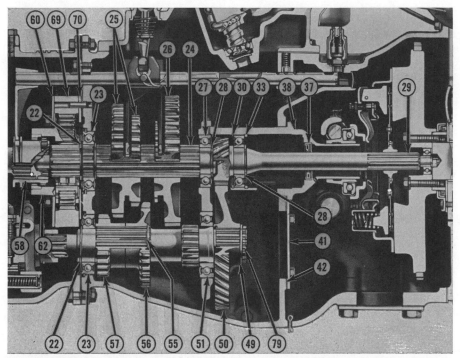

Fig. MF353—Side sectional view of typical single plate clutch and transmission assembly. The transmission countershaft is of solid construction. Refer to legend for Fig. MF357.

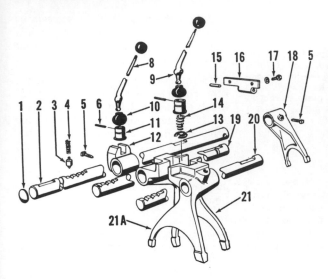

Fig. MF354 — Exploded view of shift rails and forks. Forks (21 & 21A) are interchangeable, but rails (19 & 20) are not.

1. Expansion plug
2. Planetary rail
3. Detent plunger
4. Spring
5. Fork set screw
6. Pin
8. Planetary shift lever
9. Shift lever
10. Shift lever cover
11. Shift lever cup
12. Planetary shift rail selector
13. Spring seat
14. Spring
15. Selector lock pin
16. Lever stop plate
17. Cap screw
18. Planetary shifter fork
19. Low and reverse rail
20. Intermediate and high rail
21 & 21A. Shifter forks

183. MAIN (SLIDING GEAR) SHAFT. To remove the transmission main shaft, first remove the transmission assembly (paragraph 176), main drive shaft and gear (paragraph 180 or 181) and the shifter rails and forks (paragraph 182).

Remove the four cap screws retaining the planetary unit to the transmission case and withdraw the planetary rear cover plate (60—Fig. MF356), thrust washer (61) and planet carrier (62). Using two screw drivers, work the planetary ring gear (69) and dowels from locating holes in case and remove the planetary front cover (70) and shim (71).

Remove snap ring (28—Fig. MF357) from front of mainshaft, block up between the mainshaft sliding gears and carefully bump mainshaft rearward and out of transmission case. Rear bearing can be removed from shaft and front bearing can be removed from case at this time.

When reinstalling the shaft, proceed as follows: Place the mainshaft sliding gears in position; larger gear goes to the front with the fork groove toward rear and larger gear of cluster gear goes toward rear. Install the mainshaft and rear bearing assembly. Install the bronze thrust washer (61—Fig. MF356) in recess of planetary ring gear. Install cover (70) on ring gear dowels with oil grooves of cover toward the bronze thrust washer. Install shim (71) on dowels and use a lead hammer to bump dowels into transmission case, making certain that thrust washer is in ring gear recess and is free to rotate. Install planet carrier, other thrust washer (61) and rear cover (60) with oil grooves of

cover toward the bronze thrust washer. Install and tighten cap screws retaining planetary unit to transmission case.

Remove snap ring (49—Fig. MF357) and pull the countershaft drive gear (50) from countershaft, thereby providing sufficient room to install the mainshaft front bearing. Use a suitable piece of pipe and drive the mainshaft front bearing into position and install the snap ring. Reinstall the countershaft drive gear (50) and its retaining snap ring (49).

184. COUNTERSHAFT. To remove the countershaft, first remove the mainshaft as outlined in paragraph 183 and proceed as follows: Remove snap ring (22—Fig. MF357) from rear end of countershaft and snap ring (49) and gear (50) from front end of shaft. Carefully bump the countershaft forward and out of case. Front bearing (51) can be removed from shaft and rear bearing (23) can be removed from case at this time. On models with live power take-off, renew the needle bearing (54) in rear end of shaft if bearing is damaged.

When reassembling, install the countershaft gears with the small gear to rear and with long hubs of gears together. Push the countershaft in as far as it will go and wedge a wooden block in the front compartment of the transmission housing to prevent the countershaft from moving forward while rear bearing is being drifted on. Using a suitable piece of pipe, drive the countershaft rear bearing into position, install the snap ring and remove the wooden wedge.

185. REVERSE IDLER. The reverse idler gear and shaft can be removed after removing the mainshaft as outlined in paragraph 183 and either be-

fore or after removing the countershaft. Remove cap screw and stop (77—Fig. MF357) and pull the idler shaft rearward.

The reverse idler gear contains two needle bearings which can be renewed at this time. Install the reverse idler gear cluster with larger gear to rear and with a bronze thrust washer (76) on each side of gear.

186. PLANETARY UNIT. The planetary unit can be removed after detaching the transmission from the rear axle center housing; or, if the planetary unit is the only unit to be serviced, it can be removed through the top of the rear axle center housing as follows:

Remove the hydraulic lift cover as outlined in paragraph 218 and proceed as follows:

Working through the top opening in the rear axle center housing, remove the large cotter pin, collapse and remove the rear drive shaft assembly. Remove set screw (5 — Fig. MF354) and withdraw the planetary shifter fork and coupler. Remove the four cap screws (59—Fig. MF357) retaining the plentary unit to the transmission case and withdraw the planetary rear cover plate (60), thrust washer (61) and planet carrier (62). Using

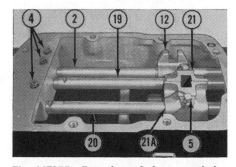

Fig. MF355—Top view of the transmission showing the installation of the shifter rails and forks. Make certain that springs (4) are in position before installing the top cover. Refer to Fig. MF354 for legend.

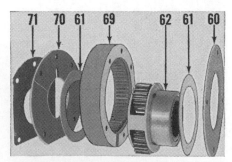

Fig. MF356 — Partially exploded view of planetary unit. Refer to Fig. MF357 for legend.

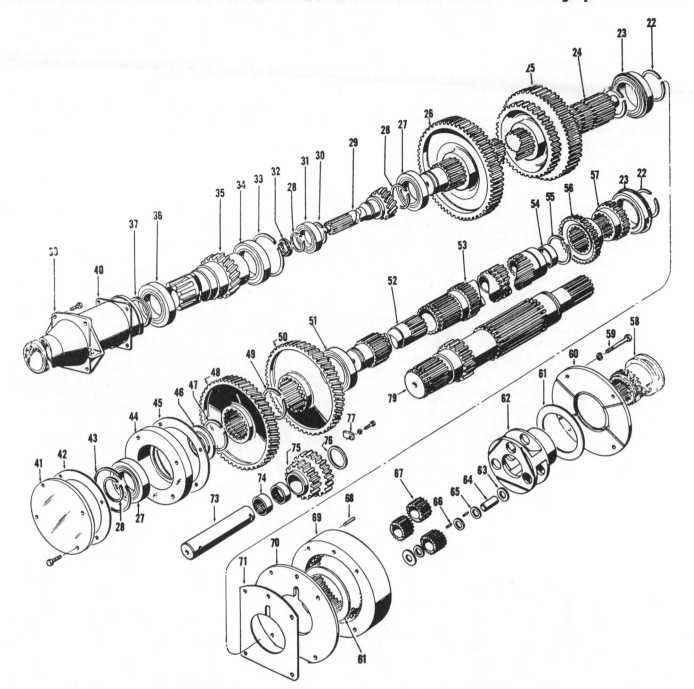

Fig. MF357—Exploded view of transmission shafts and gears. Countershaft (53), used on dual clutch models, is hollow to accommodate the pto front main drive shaft (52). Countershaft (79), used on other models is solid.

22. Snap ring	32. Oil seal*	41. Cap	50. Countershaft drive	58. Planetary shift	68. Dowel pin

22. Snap ring
23. Ball bearing
24. Mainshaft
25. Intermediate and high sliding gear.
26. Low speed sliding gear.
27. Ball bearing
28. Snap ring
29. Main drive shaft pinion.
30. Snap ring
31. Ball bearing

32. Oil seal*
33. Snap ring
34. Ball bearing*
35. Pto main drive shaft pinion*
36. Ball bearing*
37. Oil seal
38. Main drive shaft retainer
40. Gasket

41. Cap
42. Gasket
43. Snap ring
44. Pto shaft front bearing housing*
45. Gasket*
46. Thrust washer*
47. Snap ring*
48. Pto main drive gear*
49. Snap ring

50. Countershaft drive gear
51. Ball bearing
52. Pto front main drive shaft*
53. Countershaft*
54. Needle bearing*
56. High speed pinion
57. Intermediate speed pinion

58. Planetary shift coupler
59. Cap screw
60. Planetary rear cover plate
61. Thrust washer
62. Planet carrier
63. Planetary pinion thrust washer
64. Shaft
65. Washer
66. Needle rollers
67. Planetary pinion

68. Dowel pin
69. Planetary ring gear
70. Planetary front cover
71. Shim
73. Reverse shaft
74. Needle bearing
75. Reverse cluster gear
76. Washer
77. Stop
79. Solid countershaft

*Parts so marked are used only on models with live pto.

two screw drivers or equivalent, work the planetary ring gear (69) and dowels from locating holes in case and remove the front cover (70) and shim (71).

Using a suitable press and arbor, push the planetary pinion shafts (64)

out of planet carrier and remove the planetary pinions and washers.

Thoroughly clean all parts and examine them for excessive wear. Renew any questionable parts. Heavy grease will facilitate installation of the planetery pinions and rollers.

To install the planetary unit, proceed as follows: Install one of the bronze thrust washers (61) in recess of planetary ring gear. Install cover (70) on ring gear dowels with oil grooves of cover toward the bronze thrust washer. Install shim (71) on dowels and use a lead hammer to bump dowels into transmission case, making certain that thrust washer is in ring gear recess and is free to rotate. Install planet carrier, other thrust washer (61) and rear cover (60) with oil grooves of cover toward the bronze thrust washer. Install and tighten cap screws (59). Install the planetary shifter fork and coupler (58). When installing the hydraulic lift cover, refer to paragraph 218.

TRANSMISSION (Revers-O-Matic)

This section covers the torque converter, pump, clutch units, controls, and gear units of the "Revers-O-Matic" transmission installed in model MF204 Work Bull tractor.

OPERATION
Model MF204

187. The "Revers-O-Matic" drive unit consists of a pair of hydraulically actuated multiple disc clutches driven in opposite directions by a gear arrangement. The input shaft is splined to the torque converter turbine and is driven at engine speed under light loads; or, at less than engine speed at multiplied torque under heavier load. The hydraulic pump which supplies oil for the torque converter and reversing clutches is driven from the torque converter impeller, thus always rotates at engine speed. The input

shaft also serves as the pressure plate for the front multiple disc clutch assembly and, through the reverse idler, as the driving gear for the "Revers-O-Matic" countershaft. The rear gear of the countershaft is meshed with a gear attached to the pressure plate of the rear clutch; thus, at any time the input shaft is turning, the front and rear clutch housings are being driven in opposite directions at input shaft speed. Both clutch hubs are splined to the output shaft which drives the transmission countershaft. When the front clutch unit is hydraulically engaged, the output shaft, transmission countershaft and pto output shaft are all driven in the normal direction of rotation. When the rear clutch unit is hydraulically engaged the output shaft, transmission countershaft and pto output shaft are all driven opposite the normal direction of rotation. The pto driven hydraulic system pump will also be driven in reverse direction; however, due to pump design, output will not be affected. Note: If any implement or testing equipment such as a dynamometer, is being attached to the tractor power take-off shaft, this fact must be remembered. To operate stationary pto equipment, the main transmission must be in neutral and the forward actuating pedal depressed in order for the pto shaft to operate properly. Also, it is necessary that one of the directional actuating pedals be depressed when testing or operating the tractor hydraulic system.

It is very important when operating a "Revers-O-Matic" equipped tractor, that the governor low-idle speed be properly set and that the hand throttle lever be moved to the slowest speed position. Acceleration and speed control must be accomplished with the operating pedal. The clutch unit is so designed that smooth engagement is possible only at the recommended low-idle speed. An engine idle speed which is too high will result in rough, jerky starts, caused by too quick clutch engagement and excessive torque from the converter. An idle speed which is too slow, however, might result in the engine dying when the operating pedal is depressed.

TROUBLE SHOOTING
Model MF204

188. **PERFORMANCE CHECK.** The "Revers-O-Matic" operating pedal controls both the direction of travel and engine speed. Refer to Fig. MF358. When either the forward (2) or re-

verse (3) actuating pedal is depressed, the first movement rotates the pedal on axis (A), moving the "Revers-O-Matic" control valve to activate either the forward or reverse multiple disc clutch. Further depressing the pedal rotates the pedal arm at shaft (B) to open the governor. The third pedal (1) opens the governor but does not actuate either of the directional clutches.

Before attempting to check or adjust the "Revers-O-Matic" unit, first adjust the governor as outlined in paragraphs 146, 147 and 148.

With engine at operating temperature, place transmission in high gear and lock the brakes securely; then, depress the forward pedal to the wide-open position, and hold while observing the engine speed on the tractor-meter. The observed engine speed should be 1050-1150 rpm. If engine speed is lower, engine is not developing full power and a tune-up is advised. If engine speed is higher, either the torque converter or forward clutch is slipping. To determine whether the converter or forward clutch is at fault, repeat the test using the reverse pedal. Equal readings would indicate that the trouble was in torque converter, while a different reading would point to clutch malfunction.

Fig. MF358—"Revers-O-Matic" operating pedals, located on right side of transmission housing.

1. Accelerator pedal
2. Forward pedal
3. Reverse pedal
A. Directional pivot
B. Accelerating pivot

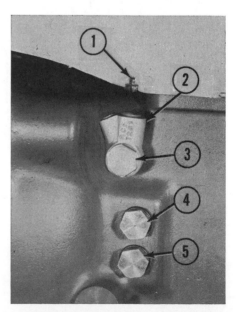

Fig. MF360 — Left side of transmission housing showing location of "Revers-O-Matic" external plugs and connections.

1. Pressure plug
2. Banjo fitting
3. Outlet plug
4. High-pressure valve cap
5. Low-pressure valve cap

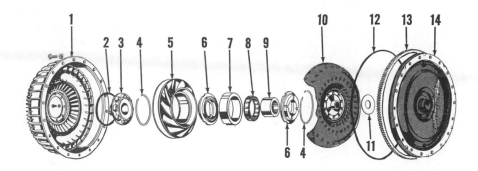

Fig. MF361—Exploded view of early torque converter unit. On late models, starter ring gear (13) is carried on flywheel rather than converter front cover (14).

1. Impeller	6. Thrust washer	10. Turbine
2. Hub seal	7. Outer clutch race	11. Thrust washer
3. Impeler hub	8. One-way	12. Sealing ring
4. Snap ring	"sprag" clutch	13. Starter ring gear
5. Stator	9. Inner race	14. Front cover

Fig. MF361A — Torque converter assembly attached to engine flywheel. Note the pump driving lugs on impeller hub. When reassembling tractor, check impeller hub for run-out with dial indicator. Runout should not exceed 0.020.

Fig. MF361B — Front view of transmission housing showing "Revers-O-Matic" oil pump, input shaft and distributor assembly.

1. Control valve linkage	3. Pump mounting bolts
2. Pressure line	4. Distributor mounting bolts
	5. Stand pipe

189. **PRESSURE CHECK.** To check the control oil pressure install a 0–300 psi pressure gage in the fitting (1—Fig. MF360) at the top of transmission housing. Set the hand throttle to obtain an engine speed of 1500 rpm, and observe gage reading. Gage pressure should be at least 150 psi. Set hand throttle to obtain an engine speed of 1800 rpm and engage each clutch in turn by depressing the pedals only enough to activate the control valve without changing engine speed. Pressure should drop to 100 psi momentarily, then return to within 5 psi of neutral reading. Failure to respond as indicated would indicate leakage in piston or shaft seals, gasket or "O" rings, or a worn pump.

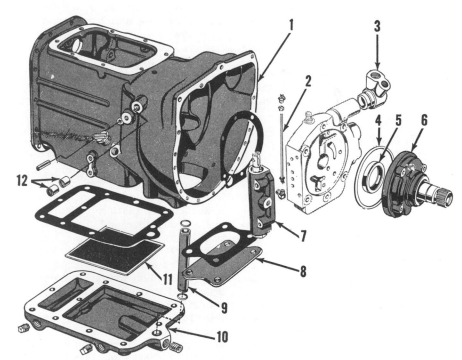

Fig. MF362 — Exploded view of transmission housing used on MF204 tractor. External dimensions are identical to housing used on MF202 and MF35 tractors.

1. Housing	5. Pump shaft seal	9. Stand pipe
2. Pressure line	6. Oil pump	10. Oil pan
3. Banjo fitting	7. Control valve	11. Oil screen
4. Gasket	8. Inspection plate	12. Needle bearing

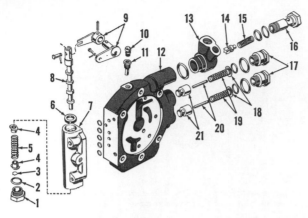

Fig. MF363 — Exploded view of control valve and oil distributor.

1. Valve cap	8. Valve spool	15. Spring
2. "O" ring	9. Valve linkage	16. Outlet plug
3. Snap ring	10. Breather	17. Valve plugs
4. Spring guide	11. Adapter	18. Shim washers
5. Centering spring	12. Distributor	19. Valve springs
6. Oil seal	13. Banjo fitting	20. Guide pins
7. Valve housing	14. Regulator valve	21. Regulating valves

Access to the two regulating valves is made through the two plugs located on the left side of transmission housing as shown in Fig. MF360. After checking the combined pressure of the two regulating valves as outlined above, remove upper plug (4) and extract the spring. Reinstall plug and check the pressure of lower regulating valve by running engine at 1800 rpm and observing the pressure on gage. Adjust pressure by adding or removing spring shims underneath regulating valve cap (5) to obtain the recommended 75-80 psi pressure. After adjusting the pressure of lower regulator, reinstall upper spring and adjust the combined pressure to the recommended 160-170 psi by adding or removing spring shims underneath the upper regulator valve cap (4).

R&R AND OVERHAUL
Model MF204

Access to torque converter, "Revers-O-Matic" pump or "Revers-O-Matic" control valve can be obtained by detaching (splitting) the transmission from the engine assembly as outlined in paragraph 190. The "Revers-O-Matic" clutch assembly can also be removed after the front split by removing the transmission oil pan, however, the component parts are more easily handled by removing the transmission case and attaching it to an engine stand, input shaft upward. To obtain access to the gear change transmission it is first necessary to remove the transmission top cover and the planetary unit as outlined in paragraph 186. Access for inspection of the transmission gears can be made without disassembly of tractor by draining the "Revers-O-Matic" unit and main transmission and removing the transmission oil pan (10—Fig. MF362).

190. SPLITTING TRACTOR. To detach engine from transmission proceed as follows: Drain cooling system, power steering system and "Revers-O-Matic" unit, remove hood and disconnect rear end of drag links. If transmission is to be overhauled, drain transmission and hydraulic reservoir. Shut off fuel and remove fuel line. Unbolt fuel tank from its rear support, loosen fuel tank front support bolts and block up between fuel tank and rocker arm cover. Disconnect heat indicator sending unit from water outlet elbow, wires from coil and generator, tractormeter cable from generator, choke rod from carburetor and oil gage line from right side of cylinder block. Loosen the U-bolt assembly from front end of throttle rod and disconnect exhaust pipe from manifold. Disconnect battery cables and remove battery. Unbolt battery platform from engine. Disconnect power steering pressure and return lines from control valve and the four power steering cylinder lines from the junction blocks under battery platform. Remove the grille front door and disconnect the two lines from oil cooling radiator. Remove the two lines to the "Revers-O-Matic" filter at left side of engine block and unbolt and remove filter and bracket. Remove the starter, then disconnect and remove the long oil cooler line. Remove one cap screw from each side of transmission to engine flange and install two aligning studs approximately 10 inches long. Support engine in a hoist and place a rolling floor jack under transmission case, unbolt engine from transmission and separate the tractor halves.

CAUTION: Roll rear end assembly

straight back, without up and down or side movement, using the two aligning studs as a guide, to avoid damage to pump driving lugs on converter housing.

When reattaching, position the drive lugs on "Revers-O-Matic" pump at right angles to lugs on converter housing. Use the two 10-inch guide studs for alignment to prevent damage to driving lugs and oil seal. Make certain that alignment is exact. After input shaft has entered converter housing, rock flywheel back and forth through 6 or 8 degrees to align splines. Do not use force. Complete the assembly by reversing disassembly procedure.

191. TORQUE CONVERTER. Power is transmitted from the engine to the drive unit through a torque converter attached to the engine flywheel. The torque converter creates a smooth, shock-free drive and multiplies torque during heavy "pull-down" loads, reducing the need for shifting gears.

The torque converter is attached to the flywheel by three drive plates, each consisting of three spring steel segments. Note: The manufacturer recommends that if drive plates are of one-piece 1/8-inch material previously used they be renewed, using the new type three-piece spring steel plates. Unbolt and remove the torque converter from the flywheel.

Note: The same standards of care and cleanliness demanded of diesel or hydraulic component servicing should be observed when overhauling the torque converter. Unless you practice good housekeeping in your shop, do not attempt to service the torque converter or components.

Fig. MF363A — Front view of "Revers-O-Matic" transmission housing with oil pump and distributor removed showing input and countershaft gears. See Fig. MF365 for legend.

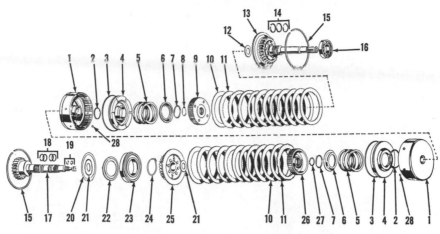

Fig. MF364—Exploded view of "Revers-O-Matic" clutch assembly.

1. Rear housing	8. Snap ring	15. Snap ring	23. Bearing
2. Piston seal	9. Clutch hub	16. Bearing	24. Snap ring
3. Piston seal	10. Steel plate	17. Output shaft	25. Pressure plate
4. Piston	11. Bronze plate	18. Sealing rings	and gear
5. Return spring	12. Thrust washer	19. Sealing rings	26. Clutch hub
6. Retainer	13. Input shaft	20. Bearing shim	27. Snap ring
7. Snap ring	14. Sealing rings	21. Thrust washer	28. Steel ball
		22. Oil slinger	

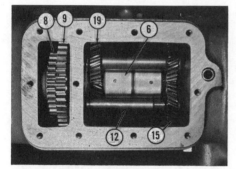

Fig. MF364A—Assembled view of "Revers-O-Matic" clutch and gear units seen from below with oil pan off. Refer to Fig. MF 365 for legend.

To disassemble the torque converter, first fashion a fixture so that the unit can be solidly supported on the rear impeller surface, with the front cover assembly up. A 2x4 frame approximately 8 inches square will serve as a suitable fixture. Place the converter in the fixture and unbolt and remove the front cover (14—Fig. MF361) and sealing "O" ring (12). Next lift off the front turbine thrust washer (11) and turbine (10). Remove the stator assembly (5), containing the one-way "sprag" clutch (7 to 9) and thrust washers (6). Remove snap ring (4) and withdraw the thrust washers and one-way clutch if service is indicated. The impeller hub (3) may be removed from impeller assembly if seal (2) needs replacing, or if the pump driving lugs are damaged. Clean the parts in a suitable solvent and renew any that are worn

or damaged. Reassemble by reversing the disassembly procedure, using new seals. Tighten the retaining bolts evenly to a torque of 22 ft.-lbs. and reinstall the torque converter on the engine flywheel. After installation, check the impeller hub for runout which should not exceed 0.020. If runout cannot be corrected by shifting the converter on the drive plates, remove the converter, check the flywheel face with a dial indicator and renew or recondition as needed.

192. **OIL PUMP.** The oil pump (6—Fig. MF362) can be unbolted and removed from distributor body after splitting tractor as outlined in paragraph 190. When reinstalling pump, note that the retaining cap screws use

Fig. MF364B — Input shaft, output gear and "Revers-O-Matic" clutch assembly.

Fig. MF364C — "Revers-O-Matic" clutch assembly with input shaft removed. Note oil tube on front end of output shaft.

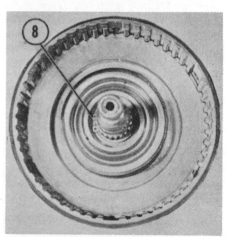

Fig. MF364D — Front end view of front clutch housing with plates and hub removed, showing snap-ring (8) which must be removed before clutch housing can be withdrawn.

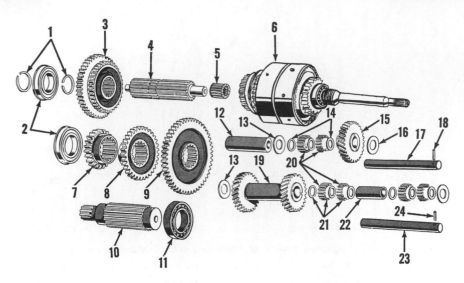

Fig. MF365—Exploded view of "Revers-O-Matic" clutch, countershaft and idler gears, and sliding gear transmission.

1. Snap ring
2. Rear bearing
3. Sliding gear
4. Main shaft
5. Pilot bearing
6. Clutch assembly
7. First gear
8. Second gear
9. Drive gear
10. Countershaft
11. Front bearing
12. Spacer
13. Thrust washer
14. Spacer washer
15. Idler gear
16. Thrust washer
17. Idler shaft
18. Roll pin
19. Countershaft gear
20. Needle bearing
21. Spacer
22. Spacer
23. Countershaft
24. Roll pin

special flat washers with bonded rubber seals. The pump shaft seal (5) and gasket (4) are the only parts which can be renewed separately, if any of the other pump parts are worn or damaged, the pump must be renewed as a unit.

193. CONTROL VALVE AND DISTRIBUTOR. To remove the control valve and distributor, first split tractor as outlined in paragraph 190 and remove oil pump. Remove oil pan (10—Fig. MF362) and withdraw stand pipe (9). Remove pressure line (2), which leads to external pressure plug. Drive the roll pin from control valve linkage (9—Fig. MF363) and withdraw linkage, then remove the three plugs (16 and 17), remove banjo fitting (13) and withdraw regulating valve springs (19) and plungers (20) from housing. Unbolt and remove distributor and control valve assembly being careful not to damage the cast iron sealing rings on input shaft. The control valve spool and housing are serviced as an assembly only, however seal (6), centering spring (5) and guides (4) are serviced. Always renew "O" rings between valve and distributor housing when reinstalling valve.

194. "REVERS-O-MATIC" CLUTCH ASSEMBLY. Note: The clutch assembly can be removed from the front of transmission after splitting tractor as outlined in paragraph 190, and removing the oil pump and distributor hous-

ing. Access and components handling are improved however, by removing the complete transmission case and attaching same to an engine stand with front end of transmission up. This latter method is recommended by the manufacturer.

In either case the disassembly procedure is as follows:

Withdraw idler gear shaft (17—Fig. MF365), idler gear (15) and spacer (12) from housing. Be careful not to loose needle bearings (20) and spacers (14) from inside the idler gear. Withdraw countershaft (23) and gear (19), then, withdraw clutch assembly (6) forward out of housing. If transmission is in a vertical position, slide countershaft gear carefully out on a light piece of tin to avoid losing the loose needle bearings (20) and spacer (22).

Place the clutch assembly, input shaft up, on a clean bench and remove snap ring (15—Fig. MF364) and withdraw input shaft and gear assembly, being careful not to damage the sealing rings (19) and oil tube on forward end of output shaft (17). If input shaft is allowed to tip, it could bend or break the output shaft oil tube, making renewal necessary. Lift off thrust washer (12), hub (9), and withdraw the clutch plates from forward clutch assembly. Remove snap ring (8), retaining the forward clutch cylinder and housing assembly (1)

and withdraw the clutch housing from output shaft. Piston (4), return spring (5) and retainers and seals will be withdrawn with the cylinder and housing assembly. To remove the piston, place clutch housing in a suitable press and, using a straddle-mounted fixture, depress spring retainer (6) and remove snap ring (7). Remove piston from its cylinder with compressed air, or by bumping open end of housing on a wood block. To disassemble the rear clutch, remove snap ring (15) and lift off clutch housing (1) containing the piston and return spring. To remove hub (26) and pressure plate (25), first remove snap ring (27) from output shaft.

Assembly is the reverse of the disassembly procedure. Internal and external splined clutch plates should be installed alternately, starting with an external spline steel plate at piston end. The missing teeth in plates should be aligned with the drain holes in housing and clutch hub.

When reassembling, temporarily install clutch assembly and distributor in the transmission housing and measure the end play of the input shaft. Vary the number of 0.015 shims (20) at the rear of output shaft bearing to obtain the recommended 0.015-0.030 end play. When the shim pack thickness has been established, remove distributor and reinstall countershaft and idler gear.

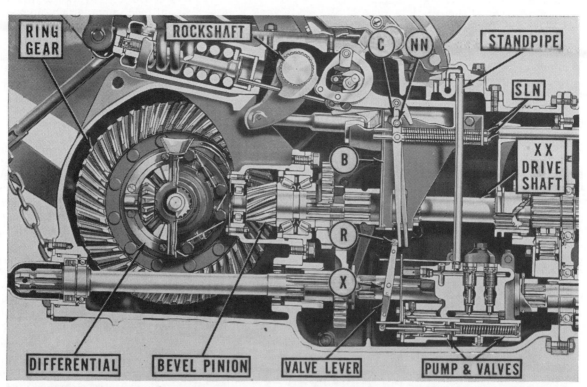

Fig. MF371—Typical sectional view of center housing, showing the installation of the internal components such as the differential, bevel pinion and hydraulic units. The ground speed pto gears are not used on some models.

195. GEAR TRANSMISSION. To inspect the transmission gears, first drain transmission and hydraulic reservoir and "Revers-O-Matic" unit and unbolt and remove oil pan (10—Fig. MF362). All transmission gears, including those of the "Revers-O-Matic" unit can then be inspected from underneath without further disassembly.

To remove the transmission gears, first split tractor between transmission and rear axle center housing, remove transmission top cover and steering gear as outlined in paragraph 14, and planetary unit as outlined in paragraph 186.

Unbolt and remove the shifter rail stop plate at the rear of the transmission case, loosen the set screws in the shifter fork and planetary selector and withdraw the shifter rails and forks. The transmission main shaft (4—Fig. MF365) can now be withdrawn from the rear of the housing. Lift the sliding gear (3) from the top of housing as the shaft is withdrawn. After removing the main shaft and gear, withdraw countershaft (10) from rear of housing and lift out gears (7, 8 and 9) from top. Reassemble by reversing the disassembly procedure.

DIFFERENTIAL, BEVEL GEARS
& REAR AXLE

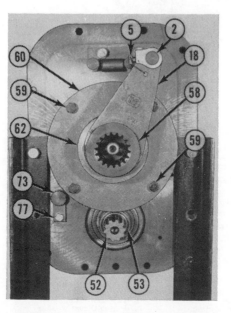

Fig. MF371A — Rear view of transmission showing the planetary unit installation.

DIFFERENTIAL

All Models

196. REMOVE & REINSTALL. The ring gear and differential unit can be removed from the rear axle center housing without disturbing the transmission or power lift unit, after removing both rear wheel and tire units, and the left rear axle shaft and housing assembly.

To remove the left axle shaft and housing unit, drain and block up the rear axle center housing. Remove left rear wheel and fender. Disconnect master brake pedal linkage at the brake shaft. Unbolt and withdraw the left rear axle shaft and housing as a unit from tractor.

When reinstalling axle shaft and housing unit, use only one standard gasket between axle housing and center housing so as to obtain correct preload on the bearings. Standard gaskets are .009-.012 thick.

197. OVERHAUL. To disassemble the differential unit, first place correlation marks on both halves of the differential case to insure correct reassembly. Remove the eight retaining bolts and separate the case halves. Differential pinions (18—Fig. MF372), spider (17) and side gears (16) can now be removed. Recommended backlash of 0.003-0.008 between the differential pinions and the side gears is controlled by the side gear thrust washers (20) and/or the pinion thrust washers (19).

If differential case (carrier) bearing cups require renewal, it is advisable to first remove the wheel axle shafts from their sleeve assemblies to obtain clearance for a puller.

The factory installed main drive bevel ring gear is retained to the differential case by rivets. Recommended procedure for removing the rivets is by drilling. Bolts are supplied with a new ring gear for installation. Bolts should be tightened to a torque of 83-88 Ft.-Lbs. Total ring gear runout should not exceed 0.002.

Tooth contact (mesh pattern) and backlash positions of the main drive bevel gears are fixed and non-adjustable.

MAIN DRIVE BEVEL GEARS
All Models

198. BEVEL PINION. The main drive bevel pinion is not available as a separate repair part without purchasing the main drive bevel ring gear. Mesh position and backlash of the bevel gears are not adjustable.

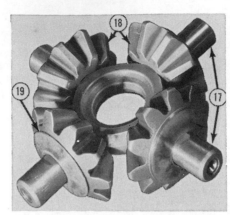

Fig. MF373—Differential spider (17), pinions (18) and pinion thrust washers (19).

To remove the bevel pinion, first remove the hydraulic lift cover as outlined in paragraph 218 and proceed as follows:

Working through the top opening in the rear axle center housing, remove the large cotter pin, collapse and remove the rear drive shaft assembly (Fig. MF371). Remove snap ring (8—Fig. MF372) and withdraw the ground speed pto drive pinion (9). Model MHF202, MF202, MF204, some MH50 and MF50 tractors are not equipped with ground speed pto gears. Remove the right wheel and tire unit. Unbolt the left rear axle housing from the rear axle center housing and pull the axle housing assembly and differential unit toward left until the bevel pinion pilot bearing (13) will clear the main drive bevel ring gear. Unbolt pinion bearing sleeve (4) from case wall and using two jack screws in the tapped holes provided in the sleeve flange, pull the pinion and sleeve assembly forward and out of case bore (Fig. MF374).

To disassemble the unit, unlock and remove nut (6-Fig. MF372) and bump pinion out of sleeve (4). If bearing cup in sleeve (4) is damaged, renew the sleeve and cup assembly as individual parts are not sold separately. When reassembling, tighten nut (6) to provide a perceptible bearing preload, or when 6-8 inch pounds of torque are required to rotate the pinion shaft in its bearings.

Install the pinion assembly by reversing the removal procedure and when installing the hydraulic lift cover, refer to paragraph 218.

199. BEVEL RING GEAR. The main drive bevel ring gear is not available as a separate repair part. If ring gear is damaged, it will be necessary to install a matched pinion and ring gear set. Mesh position and backlash of the bevel gears are not adjustable.

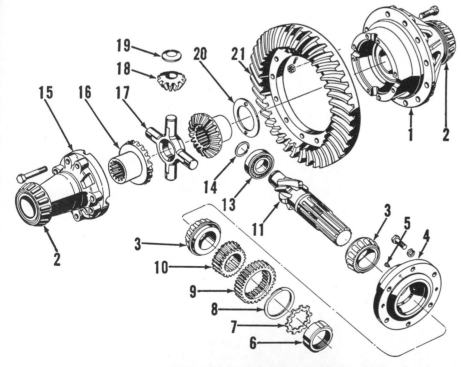

Fig. MF372—Exploded view of TO35, F40, MH50, MF35 & MF50 main drive bevel gears and differential. Mesh position and backlash of the main drive bevel gears are nonadjustable. Models MHF202, MF202 & MF204 are similar except the differential case halves are keyed.

1. Differential case, left hand	**8. Snap ring**	16. Differential side gear
2. Carrier bearing cone	9. Ground speed pto drive gear	17. Differential spider
3. Pinion bearing cone	10. Drive gear hub	18. Differential pinion
4. Pinion bearing sleeve	11. Main drive bevel pinion	19. Pinion thrust washer
5. Sleeve lock pin	13. Pinion pilot bearing	20. Side gear thrust washer
6. Lock nut	14. Snap ring	21. Main drive bevel ring gear
7. Lock washer	15. Differential case, right hand	

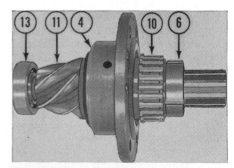

Fig. MF374—Main drive bevel pinion and bearing sleeve assembly removed from center housing. Refer to Fig. MF372 for legend.

To remove the main drive bevel ring gear, first remove the differential as outlined in paragraph 196. Factory installed main drive bevel ring gear is retained to the differential case by rivets. Recommended procedure for removing the rivets is by drilling. Bolts are supplied with a new ring gear for installation. When installing the ring gear, tighten the retaining bolts to a torque of 83-88 Ft.-Lbs. Total ring gear runout should not exceed 0.002.

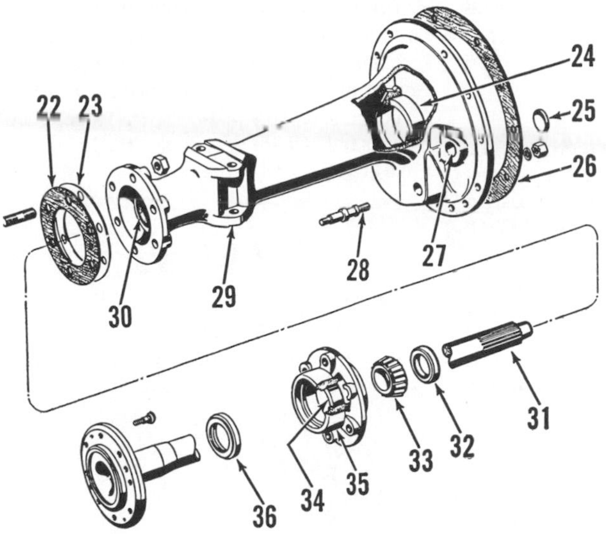

AXLE SHAFTS AND HOUSINGS
All Models

200. **BEARING ADJUSTMENT.** To check the bearing adjustment, proceed as follows: Support tractor and remove tire and wheel assemblies. Rotate either axle shaft and observe whether opposite shaft rotates in the same or opposite direction. If both axle shafts revolve in the same direction, the bearings are adjusted too tightly.

To adjust bearings, remove shims (23), shown in Fig. MF375, from between one brake back plate and one axle housing (29), until both shafts rotate in the same direction when one shaft is rotated. Then add shims (23) until the axle shafts rotate in opposite directions. This procedure will provide the recommended 0.002-0.008 clearance between inner ends of the axle shafts. Zero clearance may cause

Fig. MF376 — Exploded view of the rear wheel axle shaft and housing. Differential carrier bearing adjustment is controlled by installing one standard gasket (26) on each side of the center housing.

22. Gasket
23. Shims
24. Bearing cup
25. Expansion plug
26. Gasket
27. Brake cross shaft bushings
28. Shaft
29. Axle housing
30. Oil seal
31. Axle shaft
32. Bearing collar
33. Bearing cone
34. Bearing cup
35. Rear axle retainer
36. Oil seal

the inner ends of the axle shafts to weld together. Excessive clearance will result in damaged axle shaft oil seals. To add shims, it will be necessary to remove the axle shaft as outlined in paragraph 201.

201. **AXLE SHAFTS.** To remove an axle shaft, proceed as follows: Jack up and support both rear axle housings. Remove wheel and tire unit. Remove brake drum retaining screws, and six nuts which attach the retain-

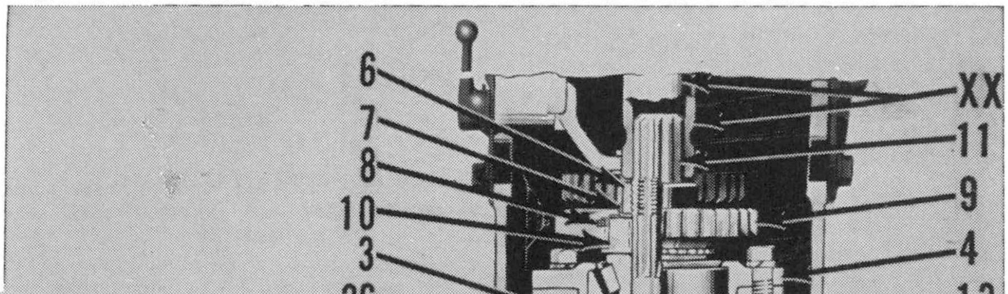

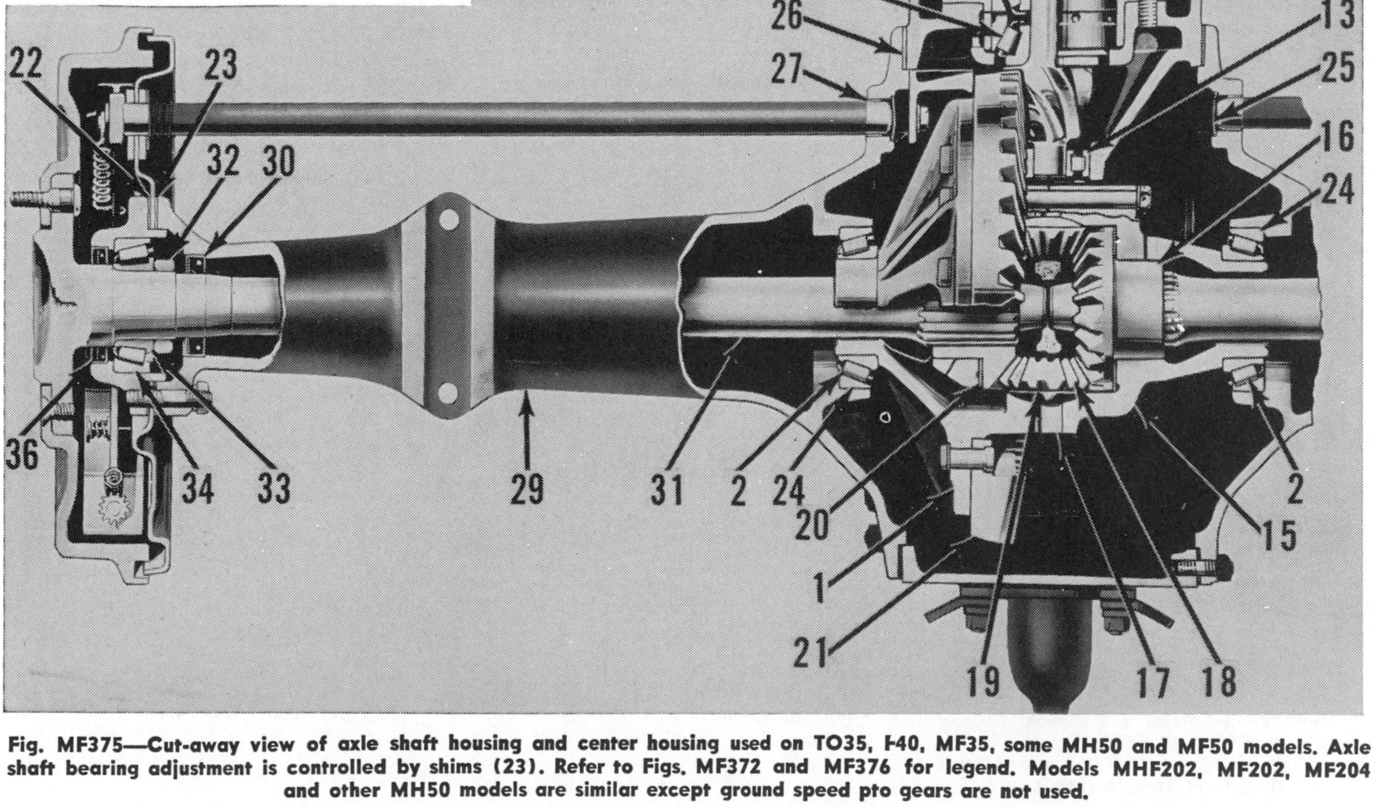

Fig. MF375—Cut-away view of axle shaft housing and center housing used on TO35, F40, MF35, some MH50 and MF50 models. Axle shaft bearing adjustment is controlled by shims (23). Refer to Figs. MF372 and MF376 for legend. Models MHF202, MF202, MF204 and other MH50 models are similar except ground speed pto gears are not used.

er (35—Fig. MF376) and the brake back plate to axle housing. Withdraw axle shaft and retainer as a unit form the housing.

202. BEARING AND OIL SEAL. To renew the inner oil seal (30) first remove the axle shaft from the housing as outlined in paragraph 201. With shaft removed, use a suitable puller or pry out the oil seal from outer end of axle housing. Install new seal with the lip facing the differential unit. To renew either the axle shaft, bearing cone and cup and/or outer oil seal, it will be necessary to remove the rear axle shaft from the axle housing, and then remove the rear axle bearing retaining collar (32—Fig. MF376). This may be accomplished by drillnig a ¼ inch hole through the steel collar and splitting same, using a cold chisel. The bearing retaining collar is a shrink fit on the axle shaft and may be pressed or driven on the shaft after heating to approximately 250 deg. F.

Install the new outer oil seal (36) with lip facing the differential unit. It will be necessary to repack the rear axle bearings with grease approximately every 1000 hours of operation. To repack the bearing, remove the axle shaft from the housing as outlined in paragraph 201.

203. SHAFT HOUSING. To remove axle shaft and housing as an assembly, first drain rear axle center housing. Block up and support rear portion of tractor. Remove fender and both rear tire and wheel assemblies. Disconnect master brake pedal linkage at the brake shaft. Remove axle housing to rear axle center housing retaining bolts and withdraw assembly.

Reinstall in reverse order of removal, using only one manufacturer's

standard thickness (0.009-0.012 new) axle housing to rear axle center housing gasket to provide the correct differential carrier bearing adjustment.

To renew the axle shaft, it will be necessary to proceed as outlined in paragraph 201.

BRAKES

All Models

204. ADJUSTMENT. Jack up and block rear portion of tractor. Remove brake adjusting port plate from inner

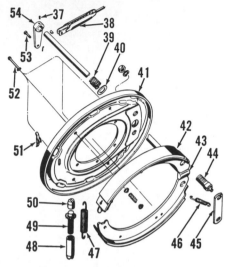

Fig. MF378—Exploded view of brake components. Brakes are adjusted with star wheel (49).

37. Set screw	47. Orange spring
38. Rod and yoke assy.	48. Adjusting screw nut
39. Spring	49. Star wheel adjust-
40. Washer	ing screw
41. Backing plate	50. Adjusting screw
42. Shoe and lining	socket
43. Actuating camshaft	51. Adjusting screw
44. Anchor	hole cover
45. Anchor pin brace	52. Shoe hold down pin
46. Blue spring	53. Clevis pin
	54. Brake lever

Fig. MF377—Adjusting the brakes. Tighten star wheel (49) until lining contacts brake drum; then, back-off the star wheel until drums are free to rotate.

face of brake back plate. Using a screw driver, turn the star wheel (49 -Fig. MF377) counter-clockwise (push screwdriver handle toward axle housing) when viewed from the back plate side, until the brake lining contacts the brake drum.

Back off the star wheel adjuster until the drums are free to rotate without any drag. Equalize the brakes by backing off the star wheel on the tight brake.

205. R&R AND OVERHAUL. The brake drum can be removed after removing the rear wheel and the brake drum to rear axle shaft flange retaining screws. Brake shoe removal is self-evident after an examination of the assembly.

An oversize (.030) brake shoe and lining kit is available as a service item. The kit is intended for use where the brake drums are resurfaced to an inside diameter of 14.030-14.035 inches.

The renewable, presized brake shaft bushing is located in the flange portion of the rear axle shaft housing and can be removed by using a suitable puller.

BELT PULLEY UNIT

All Models So Equipped

The belt pulley unit, Fig. MF379, mounted at the rear of the rear axle center housing, is a self contained drive unit and is driven by the power take-off shaft. Belt pulley rotation in either direction can be obtained by mounting the unit so that the pulley is either to the right or left of the tractor center line. Pulley can also be located in the down position.

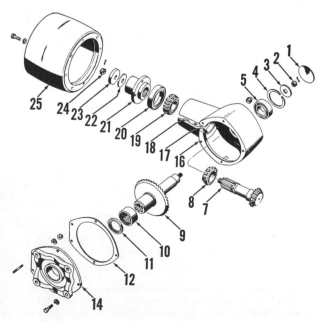

Fig. MF379 — Exploded view of belt pulley attachment.

1.	Expansion plug
2.	Nut
3.	Washer
4.	Snap ring
5.	Ball bearing
7.	Pulley drive pinion
8.	Bearing cone
9.	Pulley drive gear
10.	Needle bearing
11.	Oil seal
12.	Housing gasket
14.	Bearing housing
16.	Pulley housing
17.	Bearing cup
18.	Bearing cup
19.	Bearing cone
20.	Oil seal
21.	Pulley hub
22.	Oil seal
23.	Washer
24.	Nut
25.	Pulley

CAUTION: Never mount pulley on the right side unless a vertical muffler is installed. Never mount pulley in the "up" position, as the then top bearing will not receive proper lubrication.

206. **OVERHAUL.** Remove input shaft bearing housing (14—Fig. MF-379) and extract expansion plug (1). Remove nut (2) and bump the drive gear assembly (9) out of housing. Remove pulley, hub (21) and pinion shaft (7). The need and procedure for further disassembly is evident.

Both the driving pinion and gear are available as separate replacement parts.

When reassembling the unit, tighten nut (24) until a torque of 2-4 inch pounds is required to rotate shaft (7); then back the nut off until the cotter pin can be inserted. Tighten nut (2) until the bevel gears have a backlash of 0.004-0.006; then back the nut off until the cotter pin can be inserted.

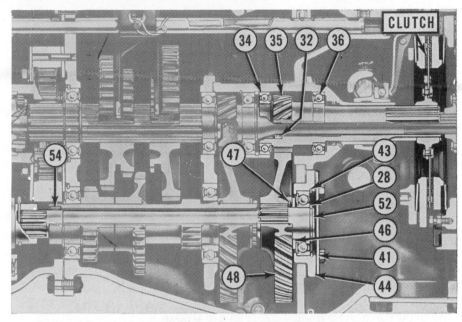

Fig. MF381—Typical sectional view of the transmission case showing the live power take-off clutch plate and driving shafts.

28. Snap ring	36. Ball bearing	47. Snap ring
32. Oil seal	41. Cap	48. Pto main drive gear
34. Ball bearing	43. Snap ring	52. Pto front main
35. Pto main drive	44. Pto shaft front	drive shaft
shaft pinion	bearing housing	54. Needle bearing
	46. Thrust washer	

POWER TAKE-OFF

OUTPUT SHAFT
All Models

207. To remove the pto output shaft, unbolt rear bearing retainer (8—Fig. MF380) from center housing and withdraw the shaft, seal and rear bearing assembly. Rear bearing (3) and seal (5) can be renewed at this time. The front needle bearing (12) can be removed after pulling its retainer (11) from the bore in center housing wall.

GROUND SPEED GEARS
All Models, Except MHF202-MF202-MF204-Some MH50-Some MF50

208. To remove the ground speed drive gear, first remove the hydraulic lift cover as outlined in paragraph 218 and proceed as follows:

Working through the top opening in the rear axle center housing, remove the large cotter pin, collapse and remove the rear drive shaft assembly (XX—Fig. MF380). Remove snap ring (23) and withdraw the ground speed pto drive gear (24).

To remove the ground speed driven gear (15) after the drive pinion has been removed, proceed as follows: Remove the pto shift cover assembly from left side of center housing and remove the pto output shaft as in paragraph 207. Remove the hydraulic pump locating dowel pins (one on each side of center housing) and withdraw the pump from the center housing. Remove the ground speed driven gear (15). Bushing (14) can be pulled from bore in sleeve (13).

When reassembling, install the hydraulic pump as in paragraph 214 or 216 and the hydraulic lift cover as in paragraph 218.

LIVE PTO MAIN DRIVE GEARS
All Models So Equipped

209. The pto main drive shaft pinion (35—Fig. MF381) is removed in conjunction with removing the transmission main drive shaft and gear (clutch shaft) as outlined in paragraphs 179 and 180.

To remove the pto main drive shaft gear (48) after the drive pinion is out, first remove the transmission shifter rails and forks as outlined in paragraph 182. The gear (48) can then be removed through top opening in transmission housing.

Fig. MF380—Sectional view of the rear axle center housing showing the pto output shaft and the ground speed gears installation. This view is typical of all models except MHF202, MF202, MF204, some MH50 and some MF50 models which do not use the ground speed pto gears (15 and 24).

XX. Driving coupling	6. Seal retainer	14. Bushing
shaft	7. "O" ring	15. Ground speed
1. Pto shaft	8. Bearing retainer	driven gear
2. Snap ring	10. Cap	16. Shift fork
3. Ball bearing	11. Bearing retainer	23. Snap ring
4. Snap ring	12. Needle bearing	24. Ground speed
5. Seal	13. Sleeve	drive gear

LIVE PTO MAIN DRIVE SHAFT
All Models So Equipped

210. To remove the live pto main drive shaft (52—Fig. MF381) first detach (split) transmission housing from engine as in paragraph 174 or 175 and proceed as follows: Remove the front pto housing bearing cap (41) and extract snap ring (28) from forward end of shaft. Use two puller cap screws in the tapped holes provided in the pto shaft front bearing housing (44) and remove the housing and bearing assembly. Remove thrust washer (46). Remove the internal snap ring (47) from the pto main drive gear (48) and withdraw the main drive shaft.

Install the shaft by reversing the removal procedure and make certain that the small shoulder on thrust washer (46) is toward front.

HYDRAULIC LIFT

Most hydraulic system trouble is caused by dirt or gum deposits. The dirt may enter from the outside, or it may show up as the result of wear or partial failure of some part of the system. The presence of gummy deposits, however, usually results from inadequate fluids or from failure to drain and renew the fluid at the recommended intervals. These principles should be kept in mind when shooting trouble on the system and also when performing repair work on the pump, valves, and cylinders.

Thus, when disassembling the pump and valves unit, it is good practice generally to not remove any parts which can be thoroughly inspected while they are installed. Internal parts of the pump, valves, and cylinder, when removed, should be handled with the same care as would be accorded the parts of a Diesel pump or injector unit, and should

be soaked or manually cleaned with an approved solvent to remove gum deposits. Unless you practice good housekeeping in your shop, do not undertake the repair of hydraulic equipment.

The transmission lubricant is the operating fluid for the hydraulic system.

NOTE: When making any of the operating checks or adjustments on model MF204 tractors, it is necessary that the gear transmission be shifted to neutral and one of the "Revers-O-Matic" operating clutches activated by depressing either the forward or reverse actuating pedal. The hydraulic pump is driven from the main transmission countershaft and does not operate with the "Revers-O-Matic" unit in neuatral. When the reverse pedal is depressed, the pump will rotate in the reverse direction, however, operation of the system will not be affected.

ADJUSTMENTS
All Models

211. Whenever any malfunction of the hydraulic system is encountered and the cause is not readily apparent, it is recommended that the following adjustments be checked in the following sequence.

Detach implement from tractor and place lower lift links in their downmost position. Place the position control lever (Fig. MF382) in the "FAST" position as shown. Place the draft adjustment slide and the draft control lever together and between the punch marks on the draft control quadrant as shown. Remove the two special cap screws from the hydraulic lift lever link (Fig. MF383) and check the master control spring for end play by pushing and pulling on the clevis. If end play is present, loosen the Allen head set screw (S) in side of housing, unscrew retainer nut (N) and withdraw the master control spring assembly as shown in Fig. MF384. Remove pin (P) and turn the clevis on the

control spring plunger until all spring end play is removed and the spring is snug but can still be rotated by finger pressure. Insert Groove pin (P) and reinstall control spring assembly.

Turn the retainer nut (N—Fig. MF-383) either way as required to eliminate all spring end play when checked by pushing in and pulling out on the clevis. Tighten the Allen head set screw (S). Note: Control spring end play will exist if retainer nut (N) is too tight as well as too loose.

Remove the inspection cover from right side of the rear axle center housing and check the control valve by moving the valve lever (Fig. MF-385) back and forth. The valve should work smoothly without any binding tendency. Use a punch or similar tool and wedge top of control valve lever rearward and away from vertical levers. Loosen nut (NN—Fig. MF386) and move the eccentric cam (C) out of contact with the cam arm.

Check the adjustment of each of the vertical levers individually. Desired adjustment is when a slight amount of finger pressure is required at lower end of lever to force it, against spring pressure, against the front of the slot in lower end of lever support bracket (B). If either of the self locking nuts (SLN) are too tight, the respective lever can be placed against the front of the slot without using pressure and

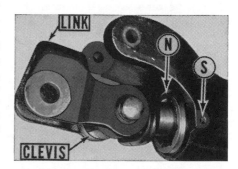

Fig. MF383—Master control spring can be checked for end play by disconnecting link from lift cover and pushing and pulling on the clevis.

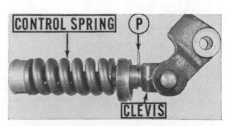

Fig. MF384 — Master control spring assembly removed from the hydraulic lift cover. Spring end play can be adjusted by turning the clevis after removing pin (P).

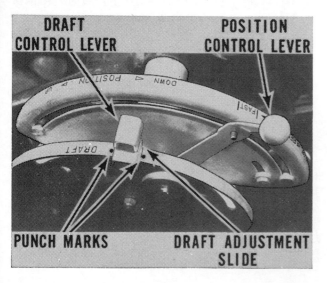

Fig. MF382 — Hydraulic system control quadrant. Control levers should be in the position shown when making any adjustments on the hydraulic system.

the tight nut should be loosened. If, on the other hand, either of the nuts are too loose, excessive pressure will be required to push the respective lever against the front of the slot and the loose nut should be tightened. Adjustments can be made through side opening in center housing by using a short ½ inch box end wrench; however some mechanics prefer to remove the lift cover before making the adjustments.

Remove the previously installed wedge and release the control valve lever. While holding the vertical levers forward, turn the small eccentric cam (C) until it firmly contacts the cam arm without moving the vertical lever and tighten nut (NN), being careful not to rotate the cam farther than desired when tightening the nut.

To check the adjustment, move the position control lever down from the "FAST" position and into the response range. As the position control lever leaves the fast position, the vertical lever should begin to move the top of the control valve lever to the rear; thereby moving the control valve toward intake.

Move the position control lever back to the "FAST" position and attach an implement (two or three bottom plow) to tractor. Turn the self-locking nut (X—Fig. MF385) either way as required so that there is just light contact between the ends of vertical levers and the roller (R) at top of the control valve lever.

This completes the internal adjustments and the inspection cover can be reinstalled on right side of center housing.

Remove the hitch pin from the rear axle center housing and insert a long piece of ¾ inch rod. Start engine and raise the lift with the position control lever until the distance between the center of the installed ¾ inch rod and the center of the pin in the lift arm is 11⅞ inches. Tighten the position control lever stop with the lever in this position. Refer to Fig. MF387.

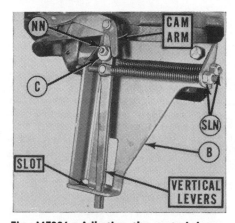

Fig. MF386—Adjusting the control levers on the hydraulic lift cover. Adjustments can be made through inspection cover opening on right side of the rear axle center housing.

Loosen the lower stop for the position control lever and carefuly move the lever down until implement begins to raise. Allow the implement to raise part way; then, raise the position control lever until the implement just drops slowly. Tighten the lower stop at this point.

Adjust the lift arm retaining cap screws so that lift arms, when raised, will just drop of their own weight.

TROUBLE SHOOTING
All Models

212. With the hydraulic system properly adjusted as outlined in paragraph 211, the following procedure should facilitate locating any malfunction.

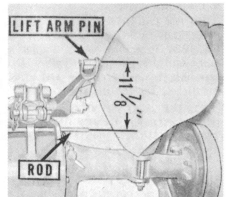

Fig. MF387 — Checking dimension (11⅞ inches) used when setting the position control lever stop.

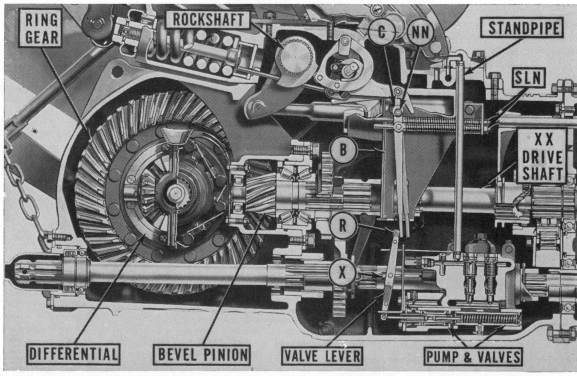

Fig. MF385—Sectional view of the rear axle center housing showing the installation of the hydraulic lift system components.

IMPLEMENT WILL NOT RAISE. Probable causes are:

a. Leak in system. Remove inspection cover and while system is operating, check for leaks at ram cylinder, stand pipe, control valve and pump side chambers.

b. Faulty relief valve. Refer to paragraph 213.

c. Broken or damaged internal pump parts. Usually indicated by a noisy pump.

d. Frozen ram cylinder. Remove side cover from center housing and see if relief valve is blowing.

IMPLEMENT LIFTS BUT WILL NOT LOWER. Probable cause is a damaged control valve spring. Remove inspection cover from center housing and determine if spring will put valve to discharge.

JERKY OR UNEVEN LIFT WHEN POSITION CONTROL LEVER IS RAISED. Probable cause is one or more faulty side chamber valves. Disassemble hydraulic pump and check for dirt or foreign material.

RELIEF VALVE BLOWS WHEN OPERATIONAL LEVER IS RAISED TO TRANSPORT POSITION. Probable causes are:

a. Check chains twisted.

b. Check chains installed in the lower holes of anchor brackets.

c. Lower links reversed.

ERRATIC ACTION OR POOR CONTROL WHEN OPERATING IN DRAFT CONTROL WITH LIGHT PRESSURE OR TENSION ON THE TOP LINK. Probable cause is damaged control linkage.

WHEN OPERATING IN POSITION CONTROL WITH KNURLED NUT SET ON QUADRANT, IMPLEMENT DOES NOT RETURN TO THE SAME POSITION WHEN THE OPERATIONAL LEVER IS RAISED AND LOWERED AGAIN TO THE STOP. Probable cause is eccentric rollers on position control link assembly.

SYSTEM OPERATING PRESSURE
All Models

213. Desired system operating pressure is 1900-2300 psi on models MH50

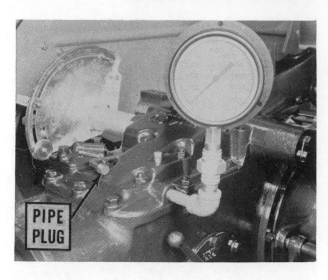

Fig. MF389 — Gage installation for checking the hydraulic system operating pressure.

Utility prior to 500630 and model TO35 prior to 168853. The recommended operating pressure on all other models is 2300-2800 psi. To identify the relief valve cartridges, refer to Fig. MF388. To check the pressure, remove the pipe plug and install a suitable pressure gage in the position shown in Fig. MF389. Start engine, move the position control lever to the raise position and note the maximum operating pressure on the gage.

If the operating pressure is too high, the relief valve cartridge should be renewed. If the operating pressure is too low, remove the inspection plate from side of center housing, operate the system and check for visible fluid leaks. If there are no visible fluid leaks, and there is turbulence around the relief valve, renew the relief valve cartridge. If the installation of a new cartridge does not bring the system operating pressure within the specified range, a faulty hydraulic pump is indicated.

PUMP
Models TO35 (Prior to 179304)-
F40-MH50—Some MF50-
MHF202-MF202

214. **REMOVE AND REINSTALL.** To remove the hydraulic pump assembly, first remove the lift cover assembly as outlined in paragraph 218. Working through the top opening in the rear

axle center housing, remove the large cotter pin, collapse and remove the rear drive shaft assembly (Fig. MF-385).

Remove the pto shift cover assembly from left side of center housing. Unbolt the pto output shaft rear bearing retainer from center housing and pull the shaft, seal and rear bearing assembly rearward. Remove the hydraulic pump locating dowel pins (one on each side of center housing) and lift the pump from tractor. Notice that the dowel pin holes in pump are of different diameter.

When reassembling, install the pump and pto shaft, then tighten the right hand dowel first so that any required shifting will automatically be made in the left dowel when it is tightened.

215. **OVERHAUL.** To disassemble the pump, unscrew the adjusting nut and

Fig. MF390—Hydraulic lift system components as viewed through front of the rear axle center housing.

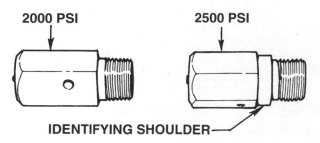

2000 PSI   2500 PSI

IDENTIFYING SHOULDER

Fig. MF388 — Two types of relief valve cartridges. The 2000 psi unit is used on some early models. Notice the identifying groove on the 2500 psi unit. Refer to text.

remove the control lever assembly (50—Fig. MF391). Remove guide (51), control valve plate (48) and control valve (43) and associated washers, "O" rings and spacers. Remove locking wire and set screw (31). Remove snap ring (21) and washer (22) and withdraw the oscillator assembly (composed of items 23 through 28). Remove retainer handle (2). Punch mark the bronze bushings (18) and right half of pump body to provide assembly marks. Unbolt and separate the two halves of pump body and remove camshaft (14), pistons (17) and blocks (16 & 37). Remove the four valve chamber plugs (3) by threading a small cap screw into plugs and pulling on the cap screw. Remove the side valve assemblies.

After pump is disassembled, thoroughly clean all parts and examine them for damage or excessive wear. Renew all "O" ring seals and any other parts which are questionable.

To reassemble the pump, proceed as follows: Reinstall the side valve assemblies and the valve chamber plugs. Position right half of pump body on bench with piston openings up and side chamber valve plugs toward you. Your left will then be the rear of the pump. Install the oscillator cam follower (15) on the shoulder of the front cam block (16) and install the front block on the camshaft with the cam follower toward rear. Install the rear cam block (37) on shaft. Follower (15) should now be between the cam blocks and on shoulder of the front block. Install pistons on cam blocks with beveled side of one piston toward front and beveled side of other piston toward rear. Install the bronze bushings (18) on camshaft with chamfered inner edge of each bushing toward cams on camshaft. Position the assembled camshaft in right side of pump body and align the previously affixed punch marks on bushings and pump body. Install the left half of the pump body and make certain that dowel pins enter holes in bushings. Tighten the four assembly cap screws to a torque of 50-55 Ft.-Lbs. Install the valve retainer handle and tighten the retaining screws to a torque of 50-55 Ft.-Lbs.

Install washer (39), "O" ring (40), retainer (41) and "O" ring (44) in control valve bore of pump body. Install washer (45), spacer (46) and washer (39) on inlet end of control valve (43) and install the assembly, making certain that the inlet slots of the control valve are in a horizontal

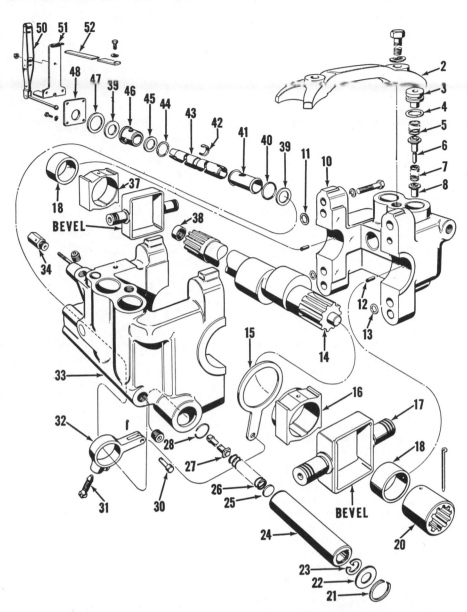

Fig. MF391—Exploded view of early type hydraulic lift system pump. Pump assembly can be removed through top opening in center housing after removing the lift cover.

plane to prevent edges of slots from contacting the floating sealing washer when the valve is returning from overload release. Install washer (47), plate (48) and guide bracket (51). Install the oscillator assembly and lock set screw (31) with wire. Install the control valve lever assembly and tighten the self-locking nut several turns.

After pump and lift cover are installed on tractor, check the system adjustments as outlined in paragraph 211.

Models TO35 (After Ser. No. 179303)—Some MF50-Some MF202-MF35-MF204

216. REMOVE AND REINSTALL. To remove the hydraulic pump assembly, first remove the lift cover assembly as outlined in paragraph 218. Work-

2. Retainer handle	27. Oscillating drive
3. Valve chamber plug	28. Snap ring
4. "O" ring	30. Clevis pin
5. Outlet valve spring	31. Set screw
6. Outlet valve	32. Oscillating link
7. Inlet valve spring	33. Body, right hand
8. Inlet valve	34. Relief (safety) valve
10. Body, left hand	37. Rear cam block
11. "O" ring	38. Needle bearing
12. Roll pin	39. Sealing washer
13. "O" ring	40. "O" ring
14. Camshaft	41. Retainer
15. Cam follower	42. Snap ring
16. Front cam block	43. Control valve
17. Pump piston	44. "O" ring
18. Bushing	45. Sealing washer
20. Coupling	46. Spacer
21. Snap ring	47. Washer
22. Washer	48. Control valve plate
23. Snap ring	50. Lever assembly
24. Oscillator body	51. Guide bracket
25. Drive retainer disc	52. Lever guide
26. Control valve spring	

ing through the top opening in the rear axle center housing, remove the large cotter pin, collapse and remove the rear drive shaft assembly (Fig. MF-385).

Remove the pto shift cover assembly from side of center housing. Unbolt the pto output shaft rear bearing retainer from center housing and pull the shaft, seal and rear bearing assembly rearward. Remove the hydraulic pump locating dowel pins (one on each side of housing) and lift pump from tractor. Notice that the dowel pin holes in pump are of different diameter.

When reassembling, install the pump and pto shaft, then tighten the right hand dowel first so that any required shifting will automatically be made in the left dowel when it is tightened.

217. OVERHAUL. To disassemble the removed pump, unscrew the adjusting nut on the control lever assembly (56—Fig. MF392). Remove guide (55), control valve plate (52), control valve (45) and associated washers, "O" rings and spacers. Remove locking wire and set screw (38). Remove snap rings (35 & 36) and withdraw oscillator assembly.

Remove cotter pin retaining front pump shaft coupling (16) to the camshaft (11) and slide coupling off. Remove cotter pin from oscillator link clevis pin (40). Remove the body bolts (9) and separate the pump body sections (2 & 3). Remove camshaft (11), pistons (13) and blocks (14 & 15).

Remove the snap rings (27); then thread a small cap screw into the valve plugs (25) and pull the plugs out. Remove the side valve assemblies (21 & 23) with their component parts.

After the pump is disassembled, thoroughly clean all parts and examine them for damage or excessive wear. Renew all "O" ring seals and any other parts which are questionable. Bearings (4 & 12) can be renewed at this time if necessary.

NOTE: Pump body needle bearings (4) have bronze bushings as an alternate for use in production and service.

To reassemble the pump, proceed as follows: Reinstall the side valve assemblies and the valve plugs. Install the oscillator cam follower (39) on the shoulder of the front block (15) and install the front block on the camshaft. Install the rear block (14) on the shaft. Follower (39) should now

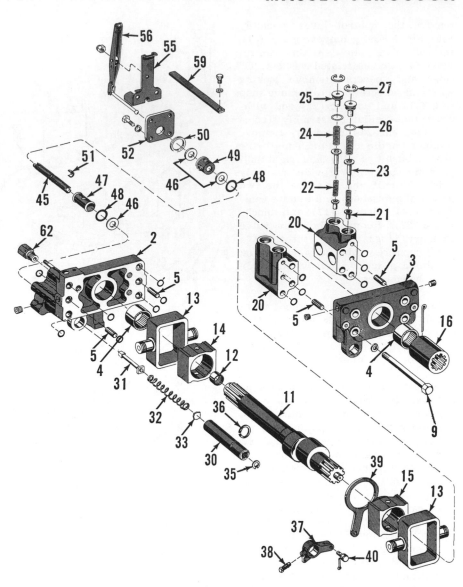

Fig. MF392—Exploded view of the late type hydraulic lift system pump. Pump assembly can be removed through top opening in center housing after removing lift cover. Note: Pump body may be equipped with bronze bushings instead of needle bearings (4). Refer to text.

2. Rear body assy.	20. Valve chamber	32. Control valve spring	47. Retainer
3. Front body assy.	21. Inlet valve	33. Retainer disc	48. "O" ring
4. Bearing (or bushing)	22. Spring	35. Snap ring	49. Spacer
5. Locating pin	23. Outlet valve	36. Snap ring	50. Washer
9. Body bolt	24. Spring	37. Oscillator link	51. Snap ring
11. Cam shaft	25. Valve plug	38. Link screw	52. Control valve plate
12. Needle bearing	26. "O" ring	39. Cam followers	55. Guide bracket
13. Block piston	27. Snap ring	40. Clevis pin	56. Valve lever
14. Rear cam block	30. Oscillator body	45. Control valve	59. Lever guide
15. Front cam block	31. Drive assembly	46. Washer	62. Safety valve
16. Front coupling			

be between the cam blocks and on shoulder of the front block. Install pistons (13) on cam blocks with machined edge of one piston facing the front body assembly and machined edge of other piston towards rear body assembly.

Slide pump body halves (2 & 3) over ends of shaft and install body bolts (9). Tighten the pump body bolts securely.

The remainder of the reassembly is self-evident.

After the pump and lift cover are installed on tractor, check the system adjustments as outlined in paragraph 211.

NOTE: If an old system (2000 psi) is converted to use the late type hydraulic pump, be sure to install a 2000 psi safety valve (62) in place of the 2500 psi valve. Difference can be noted in Fig. MF388.

LIFT COVER, WORK CYLINDER & ROCKSHAFT
All Models

218. **REMOVE AND REINSTALL ASSEMBLY.** To remove the lift cover, remove seat and disconnect upper lift links from lift arms. Remove transfer plate (Fig. MF393) from the hydraulic lift cover and withdraw the standpipe which is located directly under the transfer plate. Unbolt and remove the lift cover assembly from the rear axle center housing.

Before installing the lift cover assembly, remove the inspection cover from right side of center housing and using a punch or similar tool, wedge top of control valve lever rearward so that when lift cover is installed, the vertical levers will be in front of control valve lever.

219. **WORK CYLINDER.** With the lift cover assembly removed from tractor, the procedure for removing the work cylinder, piston and/or rings is evident. When reassembling, be sure to renew the "O" ring between the lift cover and the cylinder. See Fig. MF394.

220. **ROCKSHAFT.** The procedure for removing the rockshaft and renewing the bushings is evident. When reassembling, tighten the lift arm retaining cap screws to a point where the arms, when held in a horizontal position, will just fall of their own weight.

Fig. MF393—Before removing the hydraulic lift cover, remove transfer plate and the stand pipe under the plate.

1. Lift arm
2. Washer
3. Lock clip
4. Cap screws
5. Link shaft
6. Bushing
7. "O" ring
8. Bushing
9. Rock (lift) shaft
10. Ram arm
11. Roll pin
12. Connecting rod
13. Piston
14. Piston rings
15. "O" ring
16. Work cylinder
17. Lever support bracket
18. Spring
19. Spring guide
20. Draft control lever
21. Position control lever
22. Nut
23. Eccentric cam
24. Position control link assembly
25. Draft control pin
26. Roller cam
27. Draft control link assembly
28. Draft control rod

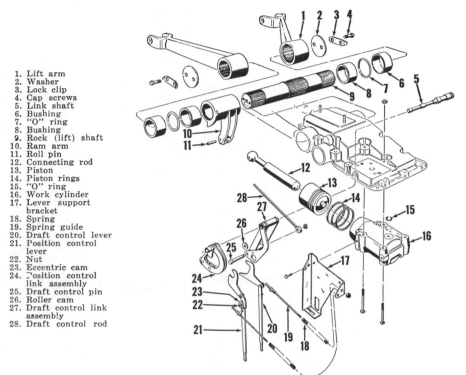

Fig. MF395—Rockshaft, work cylinder and internal control levers and linkage.

Fig. MF394—Bottom side of the hydraulic lift cover with the work cylinder removed. Note the small "O" ring located between cylinder and cover.

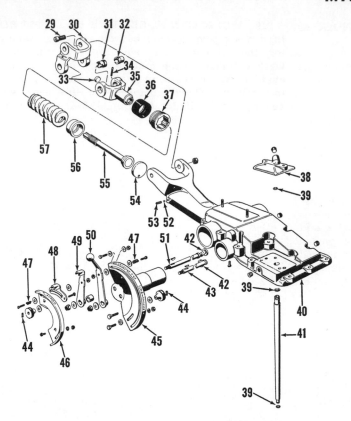

Fig. MF396—Master control spring and external control linkage exploded from the hydraulic lift cover.

29. Special cap screw
30. Rocker link
31. Bushing, right
32. Bushing, left
33. Rubber plug
34. Groov pin
35. Clevis
36. Rubber boot
37. Retainer nut

38. Transfer plate
39. "O" ring
40. Lift cover
41. Stand pipe
42. Roller
43. Position control shaft
44. Roll pin
45. Position control quadrant

46. Draft control quadrant
47. Compression spring
48. Draft adjustment slide
49. Draft control lever
50. Position control lever

51. Draft control shaft
52. Lead ball
53. Set screw
54. Overload stop disc
55. Control spring plunger
56. Control spring seat
57. Master control spring

NOTES

NOTES

NOTES